TECHNIQUES AND INSTRUMENTATION IN ANALYTICAL CHEMISTRY — VOLUME 6

ANALYSIS OF NEUROPEPTIDES BY LIQUID CHROMATOGRAPHY AND MASS SPECTROMETRY

TECHNIQUES AND INSTRUMENTATION IN ANALYTICAL CHEMISTRY

TECHNIQUES AND INSTRUMENTATION IN ANALYTICAL CHEMISTRY — VOLUME 6

ANALYSIS OF NEUROPEPTIDES BY LIQUID CHROMATOGRAPHY AND MASS SPECTROMETRY

Dominic M. Desiderio
Department of Neurology and Charles B. Stout Neuroscience Mass Spectrometry Laboratory, University of Tennessee Center for the Health Sciences, Memphis, TN 38163, U.S.A.

ELSEVIER
Amsterdam — Oxford — New York — Tokyo   1984

ELSEVIER SCIENCE PUBLISHERS B.V.
Molenwerf 1
P.O. Box 211, 1000 AE Amsterdam, The Netherlands

Distributors for the United States and Canada:

ELSEVIER SCIENCE PUBLISHING COMPANY INC.
52, Vanderbilt Avenue
New York, NY 10017

ISBN 0-444-42418-0 (Vol. 6)
ISBN 0-444-41744-3 (Series)

Printed in The Netherlands

CONTENTS

LIST OF ABBREVIATIONS USED IN THIS BOOK

a-	atto- (10^{-18})
$\overset{o}{A}$	angstrom
Ab	antibody
ACTH	adrenocorticotropin hormone
Ag	antigen
Ag*	radioactive antigen
amino acids	(A = ala; C = cys; D = asp; E = glu;
	F = phe; G = gly; H = his; I = ile;
	K = lys; L = leu; M = met; N = asn;
	P = pro; Q = gln; R = arg; S = ser;
	T = thr; V = val; W = trp; Y = tyr)
a.m.u.	atomic mass units
API	atmospheric pressure ionization
AUFS	absorbance units full scale
B	magnetic field
B_{max}	maximum binding
BA	bioassay
B/E	ratio of magnetic-to-electric fields
B/F	ratio of bound-to-free
C_{18}	octadecyl
CAD	collision activated dissociation,
	collisionally activated dissociation.
	Also, collision(ally) induced
	dissociations. Also, instead of
	dissociations, decompositions may be
	used
CCK	cholecystokinin
CI	chemical ionization
CLIP	corticotropin-like intermediate peptide
CNS	central nervous sytem
CRF	corticotropin releasing factor
CSF	cerebrospinal fluid
DADI	direct analysis of daughter ions

DAP	dipeptidylaminopeptidase
DCI	direct chemical ionization
DFI	direct fluid injection
DLI	direct liquid injection
E	electric field
EI	electron ionization
f-	femto- (10^{-15})
FAB	fast atom bombardment
FD	field desorption
FFR	field-free region
fmol	femtomole
fsec	femtosecond
GABA	gamma aminobutyric acid
GC	gas chromatography
GC-MS	gas chromatography-mass spectrometry
GHz	gigahertz $(10^9 Hz)$
GI	gastrointestinal
HAc	acetic acid
HETP	height equivalent to theoretical plate
HPLC	high performance liquid chromatography
IKE(S)	ion kinetic energy (spectroscopy)
ir	immunoreactive
K_a	binding constant
KeV	kilo-electron volts
LC	liquid chromatography
LC-MS	liquid chromatography-mass spectrometry
LC-MS-MS	liquid chromatography-mass spectrometry-mass spectrometry
LE	leucine enkephalin (YGGFL)
LPH	beta lipotropic hormone
$M^{+\cdot}$	molecular ion-radical
$(M+H)^+$	protonated molecular ion
MeV	million electron volts
ME	methionine enkephalin (YGGFM)
MIKE(s)	mass-analyzed ion kinetic energy (spectroscopy)
msec	millisecond
MSH	alpha melanocyte stimulating hormone

MS/MS	mass spectrometry/mass spectrometry
	also, tandem mass spectrometry
MW	molecular weight
N-	amino end of a peptide
nl	nanoliter
nm	nanometer
NMR	nuclear magnetic resonance
^{18}O	^{18}oxygen
ODS	octadecylsilyl
p-	pico- (10^{-12})
PCA	pyrrolidone carboxylic acid
PD	plasma desorption
PGE_2	prostaglandin E_2
pmol	picomole
POMC	pro-opiomelanocortin
Q	quadrupole
QET	quasi-equilibrium theory
RF	radio frequency
RIA	radioimmunoassay
RP	reversed phase
RRA	radioreceptor assay
SFC	supercritical fluid chromatography
SIM	selected ion monitoring
SIMS	secondary ion mass spectrometry
TAME	para-toluenesulfonylarginine methyl ester
TEAF	triethylamine:formate
TEAP	triethylamine:phosphate
TPCK	tosylamide-2-phenylethyl chloromethyl ketone
TRF	thyrotropin releasing factor
Tris	2-amino-2-hydroxmethyl-1,3-propanediol
TSH	thyroid stimulating hormone
TSQ	triple stage quadrupole
UV	ultraviolet
V	ion accelerating voltage
VIP	vasoactive intestinal peptide

to Jay

PREFACE

"Such chemistries reside in the night darkness of the molecule that seems to think."

The Striders, Loren Eiseley

Loren Eiseley, quoted from "The Striders" in Notes of an Alchemist. Copyright 1972 Loren Eiseley. Used with permission of Charles Scribner's Sons.

The brain is one of the most exciting and fertile frontiers of scientific study. A significant amount of time, experimentation, money, and effort is being expended in a multi-faceted scientific research effort to unravel the complexities of the brain. Some of the areas which are being studied include: how the brain receives, processes, and sends information; how pain is dealt with in terms of its recognition and its reduction; anatomical interconnections among and within brain regions; memory – its initial recording, storage, and recall; creative thought; movement disorders; neuroendocrinology; neuropsychiatry; dementia; among many other aspects.

Such an area of difficult experimental research necessarily brings into play a wide variety of basic and clinical scientific disciplines, subdisciplines, and experimental techniques which involves biologists, medical personnel, dentists, neurologists, neurochemists, psychiatrists, psychologists, and analytical chemists. Answers sought to questions regarding the brain's functioning require elucidation of the molecular processes that operate within the various brain regions. Only by knowing several aspects of this problem, such as which molecules are involved, their changing concentrations during specific physiological events, their precursors and metabolites, and the enzymatic and regulatory steps which are involved in those transformations, can we treat the human organism in the normal and pathological states in any rational and objective manner. Underpinning all of these research efforts are appropriate analytical tools and techniques which are used to measure the molecules involved with the maximum level of accuracy, precision, sensitivity, and molecular specificity. Accuracy is demanded in these four critical experimental parameters to avoid any confusion or ambiguity in our experiments, conclusions, hypotheses, experiments, protocols, and treatments.

Three of the most explosively growing fields in science today include biologically important peptides, high performance liquid chromatography (HPLC), and mass spectrometry (MS). Because of the significant advances in the field of neurobiology, and particular neuropeptides, there now exists an increasingly critical need for dependable structural information. The timing is appropriate to review here the pertinent features of these three areas and to bring into a novel juxtaposition the mutually beneficial aspects of these areas.

This book is basically oriented towards a review of the literature, but in those cases where it is required, more in-depth references will be suggested to the reader. Papers are selected from the wide spectrum of available research, briefly reviewed, and presented here to indicate the panorama of research that is being conducted and to point the way to needed research.

An effort has been made to remain up-to-date with respect to the current scientific literature. Because the three fields being discussed represent some of the most rapidly expanding fields in science today, this goal is rather difficult to achieve. Towards resolution of this dilemma, a camera-ready copy is prepared as opposed to a typeset book.

The author thanks the following publishers for permitting me to use these copyrighted materials: Elsevier (Figures 4.5, 4.6, 4.7, 4.8, 7.3, 7.4, 7.7, 7.8, 7.9, 7.10, 7.11, 7.12, 7.13, 7.14 and Tables 7.1 and 7.2); Wiley (Figures 6.6., 6.7, 6.13, 7.5 and 7.6); Marcel Dekker (Figures 4.3, 4.4, 6.2, 6.3, 6.16, 7.15, 7.16 and extensive portions of Chapter 8); Freeman (Figure 3.2); American Association for the Advancement of Science (Table 2.3); Pergamon Press (Table 2.6); Academic Press (Figures 4.1 and 4.2); and Macmillan (Figure 6.4); and the American Association of Dental Research (Figures 5.3, 5.4, 5.5 and 5.6 and Table 5.2).

I thank many individuals for the positive impact that they have had on the preparation of this book. My thanks go to my secretaries Dianne Cubbins and Donna Stallings for the many revisions which they have cheerfully typed; to Genevieve Fridland, who read the entire manuscript several times, offered many significant suggestions, and assisted in most of the artwork preparation; to Drs. Paul Vouros, Alex Lawson, and Chhabil Dass, who read the entire manuscript and suggested revisions; to Drs. Rodger Foltz and Gene May, who read several chapters and suggested revisions; to Drs. Fred McLafferty, Herbert Budzikiewicz, and Simon Gaskell, who were kind enough to read portions of the manuscript while at a scientific meeting and to offer suggestions; and to all of the scientific collaborators, students, visiting scientists, sabbatical faculty, and co-workers in my laboratories, for their inspired scientific insights and excellent laboratory techniques.

Finally, I also want to thank my family - my wife Jay, my daughter Annette, and my son Dominic - for the help that they extended to me during the preparation of this book. We decided as a family that I would undertake the writing of this book, then they tolerated the many long hours that I found that I had to invest in its preparation. I appreciate their help.

Chapter 1
INTRODUCTION

1.1 OBJECTIVE

The objective of this book is to describe novel, recently developed analytical techniques which can be used to effectively measure peptides, where those methods have the distinct advantage of retaining maximum structural information. To exemplify the unique specificity feature of this measurement process, this book reviews relevant scientific literature and consolidates in this one volume the selected analytical methodologies which are pertinent and useful to quantify, in general, any selected class of endogenous biologically important compounds, but here, in particular, peptides. Basically, the analytical methodology to be discussed takes advantage of the state-of-the-art advancements in two instrumental areas-high performance liquid chromatography (HPLC) and mass spectrometry (MS).

Reversed phase (RP) HPLC is demonstrably extremely well-suited for rapid, high resolution chromatographic purification of endogenous peptides from biologic tissue extracts. On the other hand, MS techniques offer significant advantages in the measurement of endogenous peptides with optimal, state-of-the-art molecular specificity, and in the fast and facile determination of the amino acid sequence of an unknown bioactive peptide. The extensive methodological data published in these two analytical areas will not be reviewed completely here, but rather a pertinent and significant novel off-line combination of these two techniques will be presented later.

While peptides are the biologically important compounds focused on, the reader will readily appreciate the important fact that virtually any other compound type is also readily amenable to this type of analytical measurement. Corresponding scientific literature is available for other classes of compounds including steroids, fatty acids, phospholipids, nucleosides, tricyclic amines, drugs, organic acids, among many other compound types.

The philosophy and material of the book are aimed towards newcomers to any one of the three diverse fields of peptide research, MS, or HPLC. Experienced researchers in these respective fields have available excellent reviews, books, articles, and chapters on these individual topics.

1.2 MAIN THEME

The underlying philosophy presented here revolves around a concept which is rather conveniently yet deceptively simple to state. On one hand, if it were possible for a procedure to attain infinite analytical resolution during chromatographic separation, then no specific detector would be required. This phenomenon is rationalized readily by considering the fact that if, for example, the resolution of an HPLC column were infinitely high, then the retention time of any compound eluting from the column would correlate unequivocally with only one molecular structure. Conversely, if the molecular specificity of the chromatographic detector were infinitely high, then no preliminary chromatographic separation step would be required because that hypothetical detector would monitor and respond to a parameter which correlates to only one unique molecular structure. With such a specific chromatography detector, no need would exist to separate a target compound from a complex matrix.

The phrase "molecular specificity" is used throughout this volume and is defined here as that detector response which is specific to and correlates to only one unique compound.

The preceding theoretical considerations notwithstanding, analytical measurements are performed in the "real world" of analysis, where the chromatographic resolution and detection capabilities of most commercially available analytical components are less than infinite. However, it is interesting to note that the novel analytical technique of mass spectrometry/mass spectrometry (MS/MS) is approaching the latter situation stated above, where unambiguous molecular specificity of the detector is optimal and, in some cases, unequivocal (1).

The research results discussed here clearly signal the availability of a primary analytical standard method which has been needed and which now can be used to calibrate other analytical methods including radioimmunoassay (RIA), bioassay (BA), radioreceptor assay (RRA), and chromatography, especially for the field of peptides where MS has not been used significantly for quantification (2). For example, RIA is a sensitive and rapid analytical technique used for many compound types where the cost per analytical measurement is low and the putative molecular specificity is high. Column chromatography, and especially reverse phase high performance liquid chromatography (RP-HPLC), is being developed rapidly and to such a high degree of reproducibility, dependability, sensitivity, resolution, and speed that an increasing number of laboratories utilize this separation technique with a UV, fluorescent, or electrochemical detector, mostly alone or, in some cases in conjunction with other assay methods to isolate and/or quantify a compound. Some analysts combine HPLC and either RIA or RRA to achieve the second highest level of molecular specificity presently available for an analytical measurement.

The thesis presented throughout this work is that the combination of RP-HPLC and recent mass spectrometric developments provides the highest level of molecular specificity currently available for (nearly) unambiguous analytical measurement of a peptide (or other compound) derived from a biological extract. Both off-line and on-line combinations of HPLC and MS are available and will be discussed.

The principles involved in both HPLC (especially reversed phase) and MS are reviewed. Applications are given of these two analytical methodologies for separation and measurement of biologically important peptides with major emphasis on brain opioid peptides.

The word opioid (noun or adjective) is used throughout to describe the plant alkaloids related to morphine (and the large number of synthetic analogues that mimic their effects) and the endogenous animal peptides of the three major peptidergic families that share the common N-terminal sequence (Tyr-Gly-Gly-Phe. . .). The word opiate is used when it is wished to distinguish the first group of opioids; generally in the context of pharmacological actions of the exogenous drugs. The term opioid peptides is used to refer to the second group as a whole, usually in the context of elucidating their physiological role.

MS has been used for the past several decades in a prolific manner in extensive studies for the structural elucidation and quantitative measurement of many biologically important compounds (See Chapter 6). Neurotransmitters, amino acids, drugs, arachidonic acid metabolites (prostaglandins, leuko-trienes, epoxyeicosatrienoic acids, and hydroxyeicosatrienoic acids), steroids, metabolic profiling (organic acids, steroids, amino acids, drugs), nucleosides, bases, oligonucleotides, peptides, organometallics, among many other types have all been investigated with MS. MS offers the signal advantage of providing the maximum amount of molecular stucture information for a given limited amount (generally nanograms) of purified sample. Of all the different classes of compounds cited above, peptides represent that one class where MS has not been used as extensively for quantitative purposes it has been used for other classes of compounds. That limitation is a significant fact to realize, because, by following the literature, one perceives that we are on the threshold of a revolution in neurobiology research including clinical, biochemical, and analytical aspects of peptides and the need exists for accurate analytical methods. Concurrently, many new, significant and useful advancements are occurring in the fields of HPLC and MS. These two converging themes are elaborated upon in later chapters. It is the author's opinion that it is appropriate to combine the two areas of neurobiology and MS in a fashion not extensively utilized before. These two methods have been used in combination in the author's laboratory for the past several years. Therefore, it is timely to review these fields together and have this volume serve as a focal point for the two fields.

The concepts discussed above are schematically represented in Figure 1.1. The list of analytical methodologies now available for neurobiological research in the "Techniques" box includes MS, several ionization methods, collision activated dissociation (CAD) processes, linked-field scanning, computer techniques, high mass capabilities, HPLC, and RP methods. The areas in which developments are rapidly taking place are listed in the box entitled "Research Areas" and include neurosciences, neuropeptides, neuroendocrinology, neuropsychiatry, dentistry, pain, biology, biochemistry, RIA, and RRA. The purpose of this book is to serve as an interface between these two groups of topics. However, other appropriate fields which are not listed in Figure 1.1 will also benefit greatly from this combination of analytical methodologies.

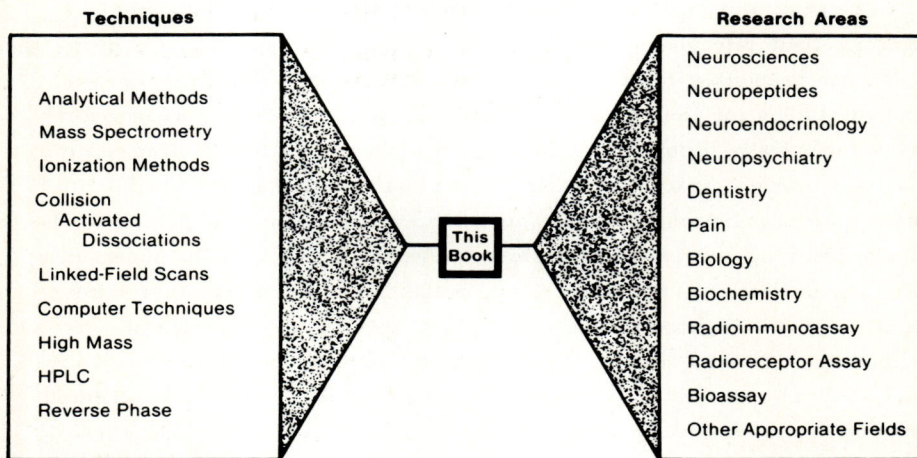

Techniques

Analytical Methods
Mass Spectrometry
Ionization Methods
Collision
    Activated
        Dissociations
Linked-Field Scans
Computer Techniques
High Mass
HPLC
Reverse Phase

This Book

Research Areas

Neurosciences
Neuropeptides
Neuroendocrinology
Neuropsychiatry
Dentistry
Pain
Biology
Biochemistry
Radioimmunoassay
Radioreceptor Assay
Bioassay
Other Appropriate Fields

Fig.1.1.   Schematic representation of the possible roles that this book can play to facilitate communication between techniques and research areas.

Therefore, as schematically represented in Figure 1.1, this book is being written to function as an interface between techniques and research areas. For example, when reading the literature available in techniques, one is aware of limited communication with research areas, and vice versa. This volume will facilitate cross-communication of workers and cross-fertilization of ideas between the two fields, and will stimulate use of the described techniques in biological neuropeptide research. The needs of one field will be focused onto the other fields. This focusing functions well in both directions and will serve both the technologist and the biologist. Also, this book will function as a referee because in some analytical techniques, namely RIA, RRA, and BA, it is imperative to have an objective standard analytical method for measurement and calibration purposes. In general, this book will interface these two highly productive areas of research which are in need of each other, to overcome the perceived barrier that, even though a vast amount of information is being published in the two areas, workers in those areas are not aware of or do not acknowledge the existence of the other. This discussion represents the rationalization for undertaking the production of this volume.

REFERENCES

1   F.W. McLafferty (Editor), Tandem Mass Spectrometry, Wiley, N.Y. (1983), 506 pp.
2   D.M. Desiderio, in J.C. Giddings, P. Grushka, J. Cazes and P.R. Brown (Editors), Adv. Chromatogr., Vol. 22, Marcel Dekker, N.Y., 1983, 1-34.

Chapter 2
NEUROPEPTIDES

2.1 INTRODUCTION

As stated in Chapter 1, the philosophy of this book is to focus attention onto the novel and powerful instrumental combination (off-line and on-line) of HPLC and MS for analysis of endogenous neuropeptides derived from a biological matrix. Because this particular combination aspect of peptide analysis is emphasized, this book will not describe completely the wide range of research which is available on peptides. Excellent reference books are available to describe the rich variety of peptide studies. The biennial American Peptide Symposium is published in book form (1-6). Basic textbooks are available on neuropeptides (7-11). Neuropeptides, neurocommunication, neurotransmission, neurochemical mechanisms, substance P in nervous system, and cerebrospinal fluid (CSF) are discussed (12-19).

Research in molecular mechanisms in memory and learning, integrators of cell and tissue function, behavior, pain, endocrinology, and neurobiology is discussed (20-25).

Extensive reviews of endogenous opioid peptides (26-29) and of the two peptides substance P and neurotensin (30) are available. A collection of papers describing the societal impact of pain is published (23).

Basic biochemistry and modulatory actions of neurotransmitters have been surveyed (31). The basic chemistry of synthesis, structure-activity relationships, and co-ordinating properties of the peptide amide bond are reviewed (32-35). Neuroanatomy (which is discussed briefly in this chapter) is reviewed in several books (36-38).

The vast biological importance and range of activities of neuropeptides, hinted to in the above brief literature survey, is not fully understood at this time. This limited understanding derives perhaps in part from the great number of mathematically possible combinations of amino acids which is available for the large number of endogenous peptides of all different lengths. It is instructive to consider the great diversity of structural, molecular, and biologic information that is contained in the linear amino acid sequence of a peptide molecule. That linear sequence plays a role in the crucial folding of a peptide/protein chain into a biologically active three-dimensional structure.

   If we assume that the linear amino acid sequence of a peptide is only one component of the information which is carried by a peptide molecule in addition to the three-dimensional information derived from both secondary and tertiary structural features, then the information content of a hexanucleotide, a six-bit computer word, and a hexapeptide may be conveniently and directly compared. A six-bit computer word has 2^6 or 64 different possible combinations, a hexanucleotide 4^6 or 4,096 different combinations, while a hexapeptide contains the far greater number of 20^6 or 64,000,000 different bits of information. These three data bases have a ratio of information content of 1:64:1,000,000, respectively.

   Biological systems have taken advantage of this great diversity, density, and flexibility of the information contained in a peptide chain as a means to effectively and efficiently provide a means to interact with stimuli. For example, Figure 2.1 is a greatly simplified schematic representation of how a cell, organelle, or organ interacts with its external environment. Some stimulus (sound, smell, touch, light, heat, etc.) arising in the outside world acts to initiate within the cell an activated chemical message. First, a ligand which is released by the action of the external stimulus interacts with a receptor to release a second messenger. While it is true that any chemical compound may serve as the first chemical messenger, the previous numerical analysis demonstrates that no class of compounds can more readily serve or offer more molecular diversity for this role than the peptide family. Indeed, different parts of a peptide molecule may serve different roles in this initial interaction process (address, message, connector, etc.). The chemical messenger, once released, will interact with a receptor in the same cell (autocrine), a neighboring cell (paracrine), or a distant cell (endocrine) to release a second messenger within that cell. Calcium ions and cyclic AMP represent two types of second messengers. These second messengers then lead to a biological event (depolarization, enzyme inhibition, etc.) which leads to a clinical manifestation (growth, death, homeostasis). It is significant to repeat the fact that the rich structural diversity of the large number of peptide families enables a variety of external stimuli to be translated with high fidelity into several internal cellular events.

   To continue the logic of this discussion, each one of the 64 million hexapeptides could conceivably be biologically unique and elicit a biologic function, activity, or further response after binding to its own specific receptor. This simple numerical consideration illustrates the large capacity for information content which is possessed by only one family of relatively short hexapeptides. This argument is readily extended to shorter and longer

```
┌─────────────────────────────────────┐
│       EXTERNAL ENVIRONMENT          │
└─────────────────────────────────────┘
                  │
                  ▼
┌─────────────────────────────────────┐
│         EXTERNAL STIMULUS           │
│   sound, smell, touch, light, heat  │
└─────────────────────────────────────┘
                  │
                  ▼
┌─────────────────────────────────────┐
│               CELL                  │
│   receptor recognition              │
│   receptor binding                  │
│   second messenger- Ca, cAMP        │
└─────────────────────────────────────┘
                  │
                  ▼
┌─────────────────────────────────────┐
│         BIOLOGICAL EVENT            │
│     depolarization,                 │
│     enzyme inhibition,              │
│     promotion, etc.                 │
└─────────────────────────────────────┘
                  │
                  ▼
┌─────────────────────────────────────┐
│       CLINICAL MANIFESTATION        │
│     growth, death, homeostasis      │
└─────────────────────────────────────┘
```

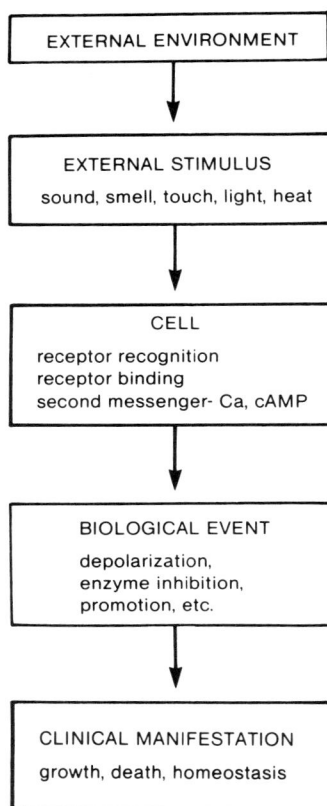

Fig.2.1. Simplified schematic representation of the interaction of a cell with an event in the external environment.

peptides. An exponential increase is realized in the complexity and information content for the longer peptides.

This chapter describes selected, biologically active neuropeptides which were discovered following the structural elucidation of the first hypothalamic releasing factor, the thyrotropin-releasing factor TRF (39-43). That discovery stimulated further research activity into the general area of brain peptides, and a number of neuropeptides were subsequently discovered and sequenced. These peptides were found to possess a wide range of biologic activities, a range that attests to the biological importance of these neuropeptides in particular and to the peptide family in general.

To provide a needed perspective to demonstrate clearly the critical need for analytical measurement methods with demonstrated unambiguous molecular specificity, a number of topics will be discussed in this chapter. These topics include: opioid peptides, neurons, peptide distribution, peptide degradation, the brain distribution of peptides, dynorphin, opioid receptors, HPLC of peptides, modified enkephalins and endorphins, urinary peptides, CSF peptides, peptides in nutrition, synthesis of peptides, neuroregulatory peptides, other neuropeptides, and concentrations of endogenous peptides. This broad range of topics is selected specifically to illustrate the crucial and pivotal role that peptides play in biological processes, the different types of peptides which are pertinent to neurochemistry, their distribution in brain structures, their chemical and biochemical synthesis, their biochemical degradation, their analysis, and the variety of metabolic roles that these peptides play. References are given to lead the interested reader to much greater detail in these areas. Because of this multifaceted nature of biological neuropeptides, there is no question regarding the need to unambiguously know the structure of the peptide which is being analytically measured in any physiological, clinical, dental, or biochemical experiment.

2.2 OPIOID PEPTIDES

The relief of pain has always been a driving need of humans. Opium, heroin, morphine, and other compounds were always sought for pain relief and/or pleasure. Once the endogenous morphine-like peptides were discovered and their chemical structures elucidated, a new level of understanding in peptide neurochemistry was attained. The structures of the endogenous substances in the brain which act as an agonist at the opioid receptor were elucidated recently (44). Two pentapeptides, named enkephalins [en kephalos, Gr., "in the head"], were isolated from porcine brain and the amino acid sequences of the two peptides were determined by employing sequential degradation techniques with a combination Dansyl-Edman procedure which indicated that the partial N-terminal tetrapeptide sequence is Tyr-Gly-Gly-Phe-. Tentative amino acid assignments indicated methionine sulfone and leucine for the fifth (C-terminal) position of the two peptides, respectively. The mass spectra of the N-acetylated, N,O-permethylated chemical derivative of the naturally-occurring enkephalins indicated that the C-terminus is methionine for the major product methionine enkephalin (ME) and (tentatively) leucine for the minor product leucine enkephalin (LE). The amino acid sequences of ME and LE, plus several of the other biologically important peptides which will be discussed in this book, are listed in Table 2.1.

The single letter amino acid code used for the amino acids in Table 2.1 is:
A = ala, C = cys, D = Asp, E = glu, F = phe, G = gly, H = his, I = ile, K =
lys, L = leu, M = met, N = asn, P = pro, Q = gln, R = arg, S = ser, T = thr,
V = val, W = trp, and Y = tyr.

The two pentapeptide sequences YGGFM and YFFFL were chemically
synthesized by classical solution methods and the mass spectra of the two
synthetic peptides were shown to be identical with the natural enkephalins.
The biologic potencies, mass spectra, and electrophoretic mobilities of these
two synthetic pentapeptides were compatible with the assigned structure of the
two naturally-occurring enkephalins.

Another naturally-occurring opioid peptide, beta-endorphin, was purified
from 200 frozen rat pituitaries and opioid activity was measured by an RRA
(45). Acidified acetone containing thiodiglycol was utilized during the
extraction procedure. These authors noted that commercial pituitary extracts
contain compounds of lower molecular weight, indicating extensive autolysis.
Autolysis of a peptide generally occurs while working with larger animals due
to the longer time required for extraction and biochemical purification of
tissues from abattoir-derived material. Pituitary beta-endorphin occurs in
conjunction with several related peptides, beta-endorphin$_{1-27}$, the des-
histidine derivative beta-endorphin$_{1-26}$ plus the corresponding N-acetylated
forms of the three endorphins. All six of these beta endorphin-related
peptides derive from one single polypeptide precursor which undergoes
differential proteolytic cleavages and acetylation reactions (46). One
processing pattern is indicated for the hypothalamus, mid-brain, and amygdala
whereas another processing pattern is observed for the hippocampus, dorsal
colliculae, and brain stem. These two processes reflect, in all probability,
the two distinct biochemical requirements of those respective brain regions.

2.3 OPIATES

Much of the research in opioids involves a study of those chemical
compounds that relieve pain, produce euphoria, ease anxiety, and facilitate
sleep. This family of natural plant alkaloid substances is known as opiates and
derives from extracts of the fist-sized top of the poppy plant Papaver
somniferm (47). "Endorphin" is a general term used to describe any
naturally-occurring opiate-like peptide and derives from combining fragments
of the two words endogenous and morphine. All endorphins described so far are
peptides. Study of opiate peptides offers the promise of producing
potentially less-addicting, more potent, and more effective pain-killers.
Morphine constitutes, by weight, approximately one-tenth of dried opium powder,

TABLE 2.1.   Amino acid sequences of biologically important peptides.

ACTH 1-39 (human)	SYSMEHFRWGKPVGKKRRPVKVYPN
Alpha-Neoendorphin	YGGFLRKYPK
Angiotensin II	DRVYIHPF
Alpha-MSH	SYSMEHFRWGKPV
Beta-Lipotropin (human)	ELAGAPPEPARDPEAPAEGAAARAELEYGLVAEAQA AEKKDEGPYKMEHFRWGSPPKDKRYGGFMTSEKS QTPLVTLFKNAIVKNAHKKGQ
Bombesin	ZQRLGNQWAVGHLM
Bradykinin	RPPGFSPFR
Corticotropin Releasing Factor (ovine)	SQEPPISLDLTFHLLREVLEMTKADQLAQQAHSNRK LLDIA-NH$_2$
Dynorphin 1-8	YGGFLRRI
Dynorphin 1-13	YGGFLRRIRPKLK
Dynorphin 1-17	YGGFLRRIRPKLKWDNQ
Dynorphin 1-32	YGGFLRRIRPKLKWDNQKRYGGFLRRQFKVVT
Dynorphin A	YGGFLRRIRPKLKWDNQ
Dynorphin B	YGGFLRRQFKVVT
Alpha Endorphin	YGGFMTSEKSQTPLVT
Beta Endorphin (1-31)	YGGFMTSEKSQTPLVTLFKNAIIKNAYKKGQ
Leucine Enkephalin	YGGFL
Methionine Enkephalin	YGGFM
Met-Enkephalin Sulfoxide	YGGFM (O)
Gastrin I (human)	ZGPWLEEEEEAYGWMDF
Insulin A (bovine)	GIVEQCCASVCSLYQLENYCN
Insulin B (bovine)	FVNQHLCGSHLVEALYLVCGERGFFYTPKA
Substance P	RPKPQQFFGLM
Arg Vasopressin	CYFQNCPRG-NH$_2$
Neurotensin	ZLYENKPRRPYIL
Oxytocin	CYIQNCPLG
Somatostatin	AGCKNFFWKTFTSC

and the amount of codeine is one-twentieth that of morphine. A popular semi-synthetic preparation is heroin which derives from acetylated morphine. Heroin is more lipophilic vis-a-vis morphine and codeine and more readily crosses the blood-brain barrier. The chemical structures of several alkaloid opiates are given in Figure 2.2.

Synthetic opiates have variable proportions of antagonist activity, where a "pure antagonist" is defined as a compound that produces none of the pharmacological actions characteristic of morphine, but can block all of morphine's effects. A common pure opiate antagonist is the compound naloxone (Figure 2.2). Other drugs such as pentazocine have great therapeutic potential and possess approximately equal proportions of agonist and antagonist activities.

Opiates exert their action through highly specific receptor sites which are protein macromolecules on the surface of neuronal cells. For example, Figure 2.3 contains a simplified schematic representation of the pertinent features of the morphine receptor (48). Several structural features are required for either alkaloid opiates or opioid peptides to interact with and bind to the receptor. Structural requirements include an anionic site covering approximately $52A^2$ with which the protonated methylamine charge interacts, a point to which charge is focused, a cavity into which a portion of the opiate molecular volume is inserted, and a flat hydrophobic where the phenyl ring interacts. Most opiate actions are stereospecific and are produced almost entirely by the (-) isomer. A bioassay of opiates is based on the fact that opiates inhibit electrically-induced contractions of smooth muscle systems (guinea pig intestine).

2.4 NEURONS

This section is necessarily brief for the purposes of this book. Much more detailed morphological, chemical, clinical, and functional information concerning the neuron is available in those books listed in Section 2.1.

Neurons are specialized cells that transmit information from one part of a biologic system to a distantly located other part and consist of a cell body which contains both dendrites and a nucleus, connected by a long portion called an axon. The nerve endings, in turn, make connection and communicate with other neurons. This highly interdigitated network of neurons is the basic mechanism that processes information in the brain. A propagation wave of ion movements is transmitted along the axon. Neurotransmitters, in general, are those chemicals which are stored in presynaptic vesicles and are released by

14

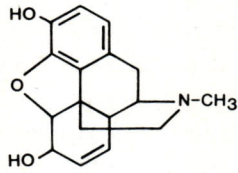

MORPHINE

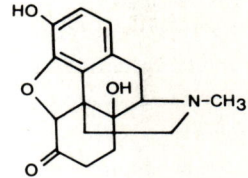

OXYMORPHONE

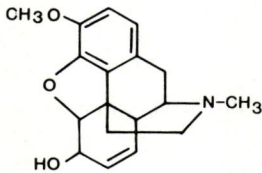

CODEINE

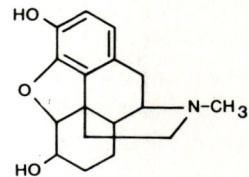

DIHYDROMORPHINE

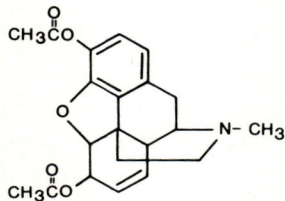

HEROIN

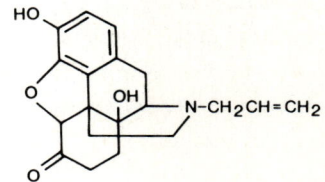

NALOXONE

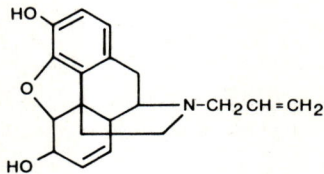

NALORPHINE

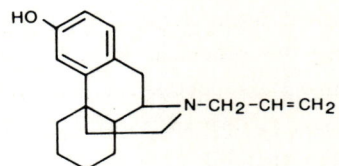

LEVALLORPHAN

Fig. 2.2.

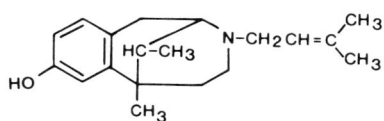

PENTAZOCINE

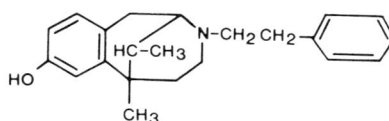

PHENZOCINE

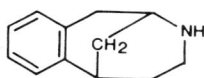

BENZOMORPHAN

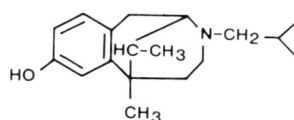

CYCLAZOCINE

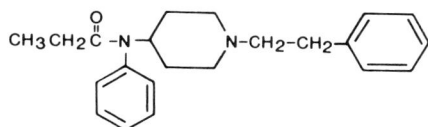

FENTANYL

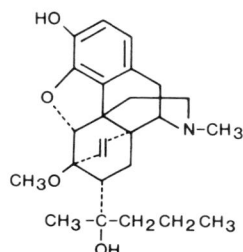

ETORPHINE

Fig. 2.2. Chemical structures of opioid compounds.

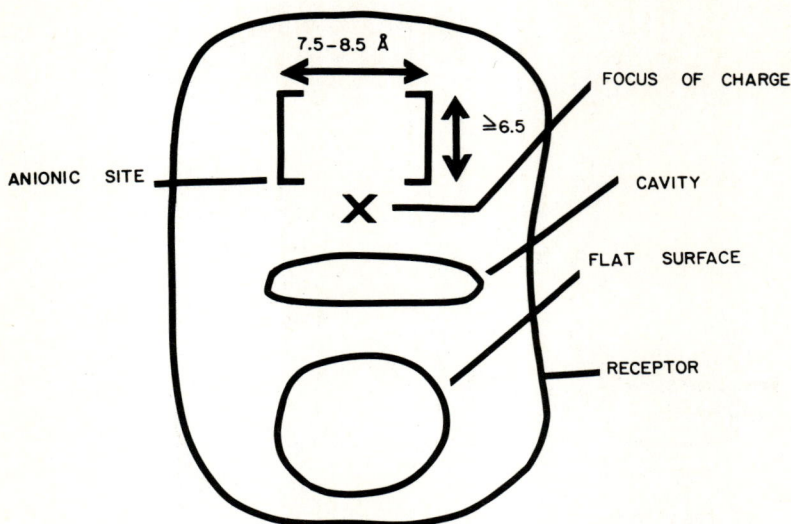

Fig.2.3. Schematic representation of the pertinent structural features of the opioid receptor.

ionic depolarization of the nerve endings of a neuron, diffuse across a non-cellular gap (approximately 200 nanometers) which is called a synapse, bind to their respective post-synaptic receptors, and influence the firing rate of neurons which are in contact with that nerve ending. Figure 2.4 contains a scheme to represent a synapse, synaptic vesicles, synaptic cleft, and a post-synaptic surface from which are elicited post-synaptic biochemical responses prompted by trans-synaptic communication by small molecules which interact with receptors. The nearby neurons, in turn, contain highly stereospecific receptor sites which are proteins embedded in the neuronal membrane and which bind those specific neurotransmitters which cross the trans-synaptic gap. It has been hypothesized that opiates exert their role by inhibiting the firing rate of certain selected neurons.

2.5 NEUROREGULATORY PEPTIDES

Neuroregulators are compounds which play a key role in communication between and amongst nerve cells. This class of compounds may be subdivided into those compounds which convey information between adjacent nerve cells (neurotransmitters) and those regulators which either amplify or decrease neuronal firing activity (neuromodulators) (49). Table 2.2 lists a variety of possible central nervous system (CNS) neuroregulators. It is

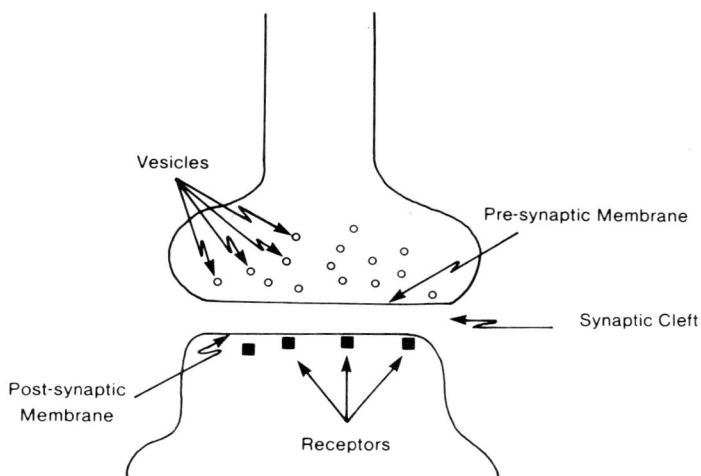

Fig.2.4. Schematic representation of a synaptic junction.

important to be able to experimentally differentiate between the two classes of neuroregulators and Table 2.3 collects those experimental criteria which are useful to distinguish neurotransmitters from neuromodulators. In a sense, the information which relates to neurotransmitters derives from experience with classical neurotransmitters, while more recent data concerning neuro-peptides forms the basis for the neuromodulators.

Classically, it was thought that only one neurotransmitter was released from a presynaptic neuron, acted with its target receptor, and then was rapidly inactivated by degradation. In contrast to this classical picture, it was found recently in a detailed study that the opiomelanotropinergic neuron system secretes at least seven peptides, which in turn may be readily biotransformed into as many as five bioactive peptides (50). These endorphin and melanocyte stimulating hormone (MSH) peptides have several biologically important activities. Both theoretical and functional analyses are needed to understand why one cell would secrete multiple transmitters. A variety of post-translational processing events of proopiomelanocortin (POMC) occurs and includes glycosylation, O-phosphorylation, acetylation, amidation, and perhaps others. This multiplicity of modifications is one of the reasons that demonstrates the need for unambiguous molecular specificity during an analytical measurement. Clearly, several different structural modifications to a peptide can provide compounds which can elicit equivalent responses from BA, RRA, or RIA.

While it is interesting to consider that MSH is almost always considered to be strictly a pigment-altering hormone, the clear understanding of the other roles that alpha-MSH play probably suffers from the misconception that these roles are limited to only those which are pigmentary in nature. A variety of extrapigmentary effects are known for alpha-MSH and includes natriuretic, lipolytic, hyperglycemic, sebotrophic, body temperature control, cardiovascular, developmental, endocrine, adrenal, thyroid, and gonadal effects.

The possibility of an extensive and intimate interaction between the endogenous opioid peptides and the classical monoamine neurotransmitters is possible as shown by light microscopic immunocytochemistry studies. The current working hypothesis in peptide neurochemistry is that the CNS contains four distinct opioid peptidergic systems: enkephalinergic, endorphinergic, substance P-ergic, and dynorphinergic and that these independent systems are available to a cell to deal with noxious and/or pleasurable stimuli. This consideration is elaborated on below.

TABLE 2.2. List of compounds which may serve as neuroregulators.

Dopamine	gamma-Hydroxybutyrate (GHB)
Norepinephrine	Glycine
Epinephrine	Taurine
Tyramine	Purine
Octopamine	Aspartate
Phenylethylamine	Glutamate
Phenylethanolamine	Corticosteroids
Dimethoxyphenylethylamine (DMPEA)	Estrogens
Tetrahydroisoquinolines	Testosterone
Serotonin (5-hydroxytryptamine)	Thyroid hormone
Melatonin	Enkephalins
Tryptamine	β-Endorphins
Dimethyltryptamine (DMT)	Substance P
5-Methoxytryptamine	Somatostatin
5-Methoxydimethyltryptamine	Angiotensin
5-hydroxydimethyltryptamine (bufotenin)	Luteinizing hormone releasing hormone (LHRH)
Tryptolines	Vasoactive intestinal polypeptide (VIP)
Acetylcholine	Adrenocorticotropic hormone (ACTH)
Histamine	Thyroid releasing hormone (TRH)
gamma-Aminobutyric acid (GABA)	Sleep factor delta

TABLE 2.3. Experimental criteria to distinguish between neurotransmitters and neuromodulators (49).

NEUROTRANSMITTER

- The substance must be present in presynaptic elements of neuronal tissue, possibly in an uneven distribution throughout the brain;
- Precursors and synthetic enzymes must be present in the neuron, usually in close proximity to the site of presumed action;
- Stimulation of afferents should cause release of the substance in physiologically significant amounts;
- Direct application of the substance to the synapse should produce responses which are identical to those of stimulating afferents.;
- There should be specific receptors present which interact with the substance; these should be in close proximity to presynaptic structures;
- Interaction of the substance with its receptor should induce changes in postsynaptic membrane permeability leading to excitatory or inhibitory postsynaptic potentials;
- Specific inactivating mechanisms should exist which stop interactions of the substance with its receptor in a physiologically reasonable time frame;
- Interventions are at postsynaptic sites or through inactivating mechanisms. The responses to stimulation of afferents or to direct application of the substance should be equal.

NEUROMODULATOR

- The substance is not acting as a neurotransmitter, in that it does not act transsynaptically;
- The substance must be present in physiological fluids and have access to the site of potential modulation in physiologically signficant concentrations;
- Alterations in endogenous concentrations of the substance should affect neuronal activity consistently and predictably;
- Direct application of the substance should mimic the effect of increasing its endogenous concentrations;
- The substance should have one or more specific sites of action through which it can alter neuronal activity;
- Interventions which alter the effects on neuronal activity of increasing endogenous concentrations of the substance should act identically when concentrations are increased by exogenous administration.

2.6 PEPTIDE DISTRIBUTION

The study of the distribution of opiate peptides in animal tissue is facilitated by fluorescent immunohistochemical techniques (51). Detailed localization studies indicate that opiate receptors and enkephalins occur in similar places, where that correspondence indicates those bodily functions which are affected by opiate drugs. For example, the substantia gelatinosa, the gray matter of the spinal cord, is packed with small neurons which interact with each other and are sensitive to nerve transmissions. Opiates are the most effective known cough suppressants and also markedly depress respiration. These visceral reflexes are regulated via the nuclei of the vagus nerve which contain high concentrations of opiate receptors and enkephalin-containing neurons. Euphoric and depressive feelings are affected by several biochemical systems in the brain such as norepinephrine, and one pathway involves cell bodies in the locus coeruleus.

Other areas which correspond to emotional responses include the brain's "limbic system". The limbic system is loosely defined to include the cingulate and parahippocampal gyri, hippocampus, septum, amygdaloid body, and hypothalamus and is believed to be a significant locus for emotion and behavioral aspects (36). Extracts of the limbic system are used (Chapter 5.4) in an RRA system to measure endogenous peptides (52). The amygdaloid nuclei constitute a prominent group of structures within the limbic system and contain the highest densities of opiate receptors and enkephalins found in the brain. Other than the intestine, enkephalins have not been detected outside the CNS. Enkephalins may play a role in the intestine by their ability to combat diarrhea and cause constipation. Opiates are also known to constrict the eye pupils and indeed, the pinpoint pupils of heroin addicts is a simple law enforcement method to assess a potential drug user. The pretectal area of the brain involved in regulating pupillary diameter has a high density of opiate receptors which may explain that observed pupil-constricting effect.

2.7 PEPTIDE DEGRADATION

When naturally-occurring enkephalins are injected into a cell, they are rapidly degraded by endogenous proteolytic enzymes. Replacing various amino acid residues of the enkephalin molecule by their stereoisomers or other amino acids greatly inhibits enzymatic activity. For example, including D-alanine at position 2, placing N-methyl phenylalanine at position 4, and modifying the structure of the methionine at position 5 by oxidation produces an enkephalin analog which is 30,000 times more potent than ME. These types of analog studies are very important because of the information which is gleaned by studying their properties versus two properties of addicting compounds - tolerance and physical dependence. Tolerance is a decreased responsiveness to a fixed dose of a drug after continuously repeating that administration, while being physical dependence causes withdrawal symptoms once the drug is no longer taken.

Peptidases located in the brain are mainly found in lysosomes which are normally structurally separated from the synaptic vessels where neuroactive peptides are presumably stored (53). While the method of animal sacrifice should not alter that separation, homogenization of tissues and/or cell disruption will probably bring products stored in these two structures into contact (see Chapter 3). By considering these experimental parameters, the importance of that irreversible inactivation of peptidases performed before tissue homogenization and analysis is recognized. Heat is a common method for enzyme inactivation. For example microwave irradiation (1.33 kilowatts, 2.45 GHz) proves to be a less satisfactory method as compared to decapitation followed by boiling of the intact tissue. The latter method indicates that no alpha-endorphin is found, and that neither a higher morphine treatment nor morphine withdrawal induces significant changes of these concentrations (54).

2.8 BRAIN DISTRIBUTION OF PEPTIDES

Contents of beta-endorphin and enkephalin vary independently from one brain region to another as measured by RIA (53). These observations support the view that beta-endorphin and enkephalin are contained within separate cellular systems in the brain and neurohypophysis, again substantiating the concept of three distinct opioid peptidergic pathways.

A comparison of the analysis of ME and LE by means of a specific RIA compared to the quantification of all endorphins by a RRA indicates that the majority of the brain regions investigated contains mostly endorphins- a maximum of 14% of LE plus ME and a minimum of 86% endorphins (55). Immunogenic conjugates of the enkephalins were made with ovalbumin, where conjugates are estimated to contain 23 ME molecules (see Figure 5.7) and five LE molecules per molecule of ovalbumin, respectively. It is believed that these molar ratios of peptide:protein are critical to an understanding of the level of molecular specificity of RIA. Striatal tissue was utilized for RRA of enkephalins. Different procedures were evaluated for their efficiency of extraction and decapitation. Tris buffer extraction is found to yield lower values of both ME and LE by RIA and endorphins by RRA, compared to the higher values which were obtained for either decapitation plus HCl extraction or microwave irradiation plus HCl extraction. Because the enkephalin molecules are relatively small, these authors rationalize that they must be coupled to the larger ovalbumin molecule to be able to raise antibodies in the rabbit, and that the putative molecular specificity of the obtained antiserum is indicated by the absence of cross-reactivity to the enkephalin metabolites. These significant structural concepts are discussed in greater detail in Chapter 5. Stereospecificity of the antiserum is presumably indicated by the low cross-reaction with Tyr-Gly-Gly-(D)Phe-Met. The authors conclude that the specificity of the antiserum is probably high because the pentapeptides retain their three-dimensional conformation, even when coupled to the antigenic protein carrier. This point is quite crucial and awaits more definitive experimental structural confirmation. That statement reminds an experimentalist that the degree to which a peptide molecule will retain its three-dimensional structure and biological activity must be considered after it is eluted from an HPLC column with an organic modifier such as acetonitrile. The preparation of the membrane fraction from rat striatum is selected because it is reasonably convenient for routine assays. It is important to carefully discard supernatants after centrifugation in order to remove a great proportion of the endogenous peptidases.

ME is suggested to be secreted into the circulation from adrenal glands (56). The majority of the immunoreactive ME is located in the medulla with a small amount from the cortex. Data suggest that the majority of the immunoreactivity in the plasma, after using extraction conditions which include inhibition of proteolysis, is located in ME (85%), with a small amount (15%) which may be due to the C-terminal tetrapeptide metabolite Gly-Gly-Phe-Met. A mean value of 22 pg ml^{-1} of ME in plasma is determined utilizing a highly specific RIA developed using methionine sulfoxide enkephalin. Therefore, it is necessary to oxidize all samples with hydrogen peroxide before analysis. These studies also include peptide measurements in CSF, where a range of 5-29 pg ml^{-1} was found. This study suggests that ME circulating in CSF is more likely to be of brain rather than of pituitary origin.

This study indicates that, during chemical iodination, the amino terminal tyrosine residue of ME is altered by the addition of iodine, and the methionine carboxy terminus is oxidized to methionine sulfoxide, thereby losing two major antigenic determinants. However, by immunizing with methionine sulfoxide enkephalin coupled by its amino terminus to bovine thyroglobulin, the antigenicity of the carboxy terminus of enkephalin is increased. It is possible to protect ME against air oxidation by use of the mild antioxidant, thiodiglycol. The molar ratio of methionine sulfoxide enkephalin to bovine thyroglobulin is 17:1.

The concept that both endorphin and ME comprise two distinct endorphinergic and enkephalinergic systems, respectively, in the brain is supported by an RIA study on the regional distribution of the two peptides in 42 discrete areas of rat brain (57). Immunohistochemical studies illustrate that the enkephalin-containing fiber distribution parallels that of opiate receptors, with the highest density in the globus pallidus. Parallels are seen between the peptide distributions in both bovine and rat brains. ME is almost always present in higher concentrations compared to LE, with a ratio ranging from one in the anterior thalamus to nine in the nucleus accumbens.

CNS homogenates contain proteolytic enzymes which may rapidly metabolize enkephalins (58). Therefore, the stability of LE was determined in rat carcasses which were maintained at 4°C in an effort to stimulate human post-mortem conditions as closely as possible. LE immunoreactivity did not display any significant change in either the cortex or hypothalamus tissue for up to eight hours, a time which approximates the mean time interval between the time of human death and the time of obtaining brain tissue at autopsy. In that study, LE is conjugated to bovine thyroglobulin (MW=660,000), and the conjugate contained approximately 25 micrograms LE mg^{-1} of protein, an amount which corresponds to a peptide: protein molar ratio of 30:1. Twenty brain structures were studied, and the highest concentrations of enkephalins are found in the pituitary, infundibular stalk, globus pallidus, putamen, substantia nigra, amygdala, caudate head, hypothalamus, and spinal cord.

Another study indicates the relatively high stability of the enkephalin and substance P neuropeptide molecules in post-mortem animal brain tissues, a fact which suggests that the content of those peptides analyzed in post-mortem human brain most probably reflects the content before death (59). This study and the previous study indicate that the post-mortem stability of peptides may be maintained adequately for a limited time, if the cell wall integrity is preserved and if peptidases remain effectively compartmentalized. In this study, the highest immunoreactivities of substance P are found in the substantia nigra-pars reticulata (1,535 pmol g^{-1} tissue) while the highest content of ME is in the lateral globus pallidus at 1,163 pmol g^{-1}. The ratio of ME to LE ranges from two to six. The regional distributions of substance P and ME was determined in post-mortem tissue derived from both normal human and Huntington's disease brains. Model experiments demonstrate that both neuropeptide immunoreactivities were stable for up to 72 hours post-mortem. In Huntington's disease, a substantial drop of over 80% in substance P content of the globus pallidus and the substantia nigra was observed in addition to a reduction of over 50% in the ME content of these areas.

A possible neuroendocrine role for ME is suggested in a study with rats which indicates the presence of increased amounts of immunoreactive ME in the anterior pituitaries of "older" rats. There is no difference between young and old rats and their hypothalamic content of ME. Increased ME in the pituitary may be related to the reduced gonadotropin secretion in the old rats (60). Intense immunohistochemical staining of the intermediate lobe of the rat pituitary is observed when an antiserum is used which is raised against synthetic dynorphin$_{1-13}$ and treated with a water-soluble carbodiimide. These results indicate that the intermediate lobe staining is due mainly to four

different derivatives of beta-endorphin with an acetyl group on the amino
terminus. This coupling mechanism used for small peptides and larger
proteins to render the peptides immunogenic must be used with extreme caution.
This study indicates that, under certain conditions, such treatment of
peptides may acetylate primary amino groups, where the source of non-
biological acetate derives from commercially available peptides (61).

A small percentage (2-3%) of the tissue stores of LE and ME is released
by a pulse of potassium ions (62). The enkephalin immunoreactivity released
from rat striatal slices consists of a single component (Sephadex G-25)
and parallels a study of substance P release. On the other hand, potassium
stimulation of hypothalamic tissue releases somatostatin and beta-endorphin
immunoreactive peptides, which then display a heterogeneous molecular weight
distribution by gel filtration. The authors indicate that HPLC may be
necessary to separate various enkephalin-like peptides which migrate as one
peak through the molecular sieves which are used in this study. These data
indicate a possible role of new enkephalins in neurotransmission.

Continuous labeling and pulse-chase techniques were employed to
elucidate the mechanisms of synthesis and secretion of multiple forms of
immunoreactive beta-endorphin by cultured dispersed rat anterior lobe cells
and the intact neurointermediate pituitary lobe (63). It was found that
intact neurointermediate lobes chemically incorporated certain radiolabeled
amino acids into four to six forms of immunoreactive beta-endorphin. Four
of those forms are compatible with authentic beta-endorphin, N-acetylated
beta-endorphin, beta-endorphin$_{1-27}$, and N-acetylated beta-endorphin$_{1-27}$. The
two radiolabeled amino acids, methionine (^{35}S) and tyrosine (^{3}H), were
utilized in this study. Triethylamine-formic acid buffer was utilized with
acetonitrile as organic modifier for a RP-HPLC gradient. In the anterior
lobe, the majority of the beta-endorphin is not processed further, but
rather, it is released as an intact molecule, while, on the other hand, it
serves as a biosynthetic intermediate compound in the neurointermediate lobe.
The authors state an important consideration that the mere detection of a
peptide in a tissue extract should not be taken as direct evidence that that
peptide represents an actual physiological product. This conclusion is an
important conceptual consideration because an intermediate, incomplete
inactivation of a pituitary enzyme during extraction or else post-mortem
autolysis may possibly produce some of the detected peptides in an
artifactual manner.

The doubly-modified peptide N-acetylated beta-endorphin$_{1-27}$, which is
acetylated at the N-terminus and a tetrapeptide is removed from the

C-terminal, represents the most abundant immunoreactive species found in the pituitary, and yet it is essentially inactive as an opioid peptide. Carboxy terminal cleavage leads to a ten-fold drop in potency while N-acetylation results in virtual elimination of opioid activity (64). Clearly, a free N-terminus is required for the peptide to bind with its biological receptor. On one hand, N-acetylation appears to be a rapid process and is found shortly after formation of beta-endorphin$_{1-27}$, whereas carboxy terminus cleavage is a relatively slow process.

A parallel diurnal variation in the concentrations of both ME and LE was observed, where pinealectomy did not affect the medial basal hypothalamic ME and luteinizing hormone releasing hormone rhythms. This information implies that a common controlling mechanism outside the pineal gland may be possible to provide those daily changes. Furthermore, the daily rhythmic patterns of ME fluctuations in the medial basal hypothalamic tissue and anterior hypothalamic pre-optic area are apparently inversely related to the diurnal changes of serum testosterone levels (65).

There may be an alteration of peripheral methionine enkephalinergic functions in the genetically hypertensive rat. Spontaneously hypertensive rats have lower immunoreactive ME concentrations compared to those levels in normotensive control rats (66). These authors indicate that ME immunoreactivity can always be detected in platelet extracts from humans, rats, and rabbits whereas no immunoreacitivity is found with an antiserum which is directed towards LE. The concentration of ME immunoreactivity in platelets is lower than in some CNS and peripheral tissues (1% of striatum, 5-10% sympathetic ganglia and intestine). Data indicate that platelet ME immunoreactivity is derived from either ME or the low molecular weight enkephalin-like peptides which are present in plasma.

The olfactory bulb provides a unique and significant research opportunity to elucidate both physiological mechanisms of ME, substance P, and somatostatin as well as potential interactions between those peptides and either gamma-aminobutyric acid (GABA) or dopamine in neurons which are defined morphologically and interact in the processing of olfactory stimuli (67). Other peptides which have been identified in the olfactory bulb, but have not yet been localized include: cholecystokinin (CCK), gastrin, insulin, vasoactive intestinal polypeptide (VIP), thyrotropin releasing hormone (TRH), vasopressin, beta-endorphin, and beta-lipotropin. The presence of such a diversity of biologically active peptides in the olfactory bulb may be related to the prominent roles which olfaction plays in the regulation of nutrition, reproduction, and affective functions and again demonstrates vividly the numerical concepts relating to the information content of peptides which are described in section 2.1.

The distribution of beta-endorphin-related peptides pertaining to rat pituitary and brain concentrations has been reviewed (46). The amino acid sequence of beta-endorphin is strongly conserved over five species (pig, camel, sheep, rat, human) where the only differences observed are ^{23}Val in pig and ^{23}Ile or ^{27}Tyr in human for His. Such an extensive conservation of the amino acid sequence and structure in these diverse species indicates that both the complete amino acid sequence and the three-dimensional structure of the beta-endorphin molecule are probably important aspects for its opiate properties. Because of the difficulties commonly experienced in the chromatography of small quantities of basic hydrophobic peptides, a wide variety of solvent systems was utilized, with acetic acid (50%) being the most satisfactory eluent. Difficulties are noted with HPLC concerning ME and its sulfoxide form because HPLC will clearly separate the two enkephalins, with the more hydrophilic sulfoxide eluting first in the RP mode. Due to the presence of several N-acetylated endorphins, the acetylating reactions appear to offer another selective metabolic control mechanism for expressing different biologic activities. Immunofluorescence experiments demonstrate the presence of a highly organized neuronal network commencing with hypothalamic cell bodies, expressing through long axons, and terminating in bundles of terminals at several defined locations. The main concentration of beta-endorphin immunoreactive peptides is in the hypothalamus where cell bodies are confined to the arcuate nucleus, medium emminence, and the ventromedial border of the third ventricle. The occurrence of nerve endings in the area of the third ventricle is a significant finding because, for example, catalepsy and analgesia occur in some rats a few minutes after injection of dynorphin$_{1-13}$ into the lateral ventricle (68).

2.9 DYNORPHIN

Dynorphin$_{1-13}$ (Table 2.1) is a peptide which displays typical opioid activity in several different assay systems where it is shown to be extraordinarily potent; 700 times more potent than LE, 200 times more than morphine, and 50 times more potent than beta$_c$- endorphin. Naloxone has less potency (1/13) in blocking this peptide compared to either normorphine or LE blockage. Dynorphin$_{1-13}$ competes variably with opiate alkaloid and peptide radioligands for stereospecific binding to rat brain membranes and, while it is not as highly potent, that reduced potency may be due to rapid degradation of the dynorphin by a membrane-bound enzyme. It is significant to realize that dynorphin$_{1-13}$ is a C-terminal octapeptide extension of LE, where that octapeptide extension -Arg-Arg-Ile-Arg-Pro-Lys-Leu-Lys contains five highly basic amino acid residues. The N-terminus of dynorphin$_{1-13}$ is necessary for biologic potency because N-acylation abolishes bioactivity. By deleting the ^{6}Arg residue, the remaining carboxy terminus is shifted in the direction of the N-terminus by 500 picometers (0.5 nm), while inserting a ^{6}Gly shifts the remaining amino acid sequence towards the C-terminus by that same distance. The loss of potency in both these dynorphin analogs demonstrates that the unique amino acid sequence in the octapeptide extension and the linear distance of the octapeptide sequence from the LE N-terminal pentapeptide are critically important structural parameters for optimal biologic potency and confirm also that the spatial relationships schematically illustrated for the opiate receptor (Figure 2.3) must be rigidly maintained within 0.5 nm.

The novel opioid octapeptide dynorphin$_{1-8}$ is proposed to exist in the intermediate pituitary of rats where the peptide is isolated by immuno-absorption to antibodies directed against porcine dynorphin$_{1-13}$ (69). The opioid octapeptide may occur at a level of almost one pmol mg^{-1} within the neurointermediate pituitary of rats.

The amino acid sequence of the opoid peptide dynorphin$_{1-17}$ which is extracted from porcine pituitary is Tyr-Gly-Gly-Phe-Leu-Arg-Arg-Ile-Arg-Pro-Lys-Leu-Lys-Trp-Asp-Asn-Gln. The full biologic activity of this heptadecapeptide resides completely in the N-terminal tridecapeptide. It is significant to note that the usually susceptible Lys^{13}-Trp^{14} bond is resistant to digestion with trypsin. While trypsin-resistant Lys bonds are rare, nevertheless, they have been reported (70). The resistance of this specific peptide bond to trypsinolysis may reflect some unique microenvironmental aspect of the secondary structure of this heptadecapeptide; one possibility is the hydrophobicity of the Trp residue. Several improvements in methodology noted in this research include the avoidance of lyophilization- where large losses of peptide had been noted, use of detergents to minimize absorption losses, monitoring peak effluents by RIA instead of BA, and use of an immunoaffinity chromatography column.

The existence of a highly specific dynorphin receptor in the guinea pig myenteric plexus and brain is known, where that receptor is presumably the same as the kappa opioid receptor with which certain benzomorphans preferentially bind. Data were presented to support the hypothesis that dynorphin is a specific ligand for the kappa subclass of opioid receptors (71). Studies were conducted with the guinea pig myenteric plexus-longitudinal muscle preparation, where comparisons were made between dynorphin$_{1-13}$ and the prototypical kappa agonist ethylketocyclazocine (Figure 2.2). The guinea pig myenteric plexus is considered to contain the mu and kappa, but not the delta, opioid receptors.

Opioid receptor subclasses are distinguished on the basis of several parameters including (i) differences in physiological effects observed with different opioids; (ii) relative potencies in sensitivities to naloxone antagonism; and (iii) membrane-binding characteristics (see section 2.10).

During the isolation of the dynorphin heptadecapeptide, neutralization of the acetic acid extract precipitated most (60-70%) of the immunoreactive dynorphin, a fact that indicates that a larger dynorphin molecule is also present in that preparation (72). HPLC purification of the larger dynorphin precursor, followed by automatic Edman degradation, yields the following amino acid sequence information corresponding to a peptide having a molecular weight of 3,986: Tyr-Gly-Gly-Phe-Leu-Arg-Arg-Ile-Arg-Pro-Lys-Leu-Lys-Trp-Asp-Asn-Gln-Lys-Arg-Tyr-Gly-Gly-Phe-Leu-Arg-Arg-Gln-Phe-Lys-Val-Val-Thr. This dynorphin$_{1-32}$ is one of the more interesting opioid peptide precursors found to date because apparently it is a precursor for several related peptides (See Figure 2.5).

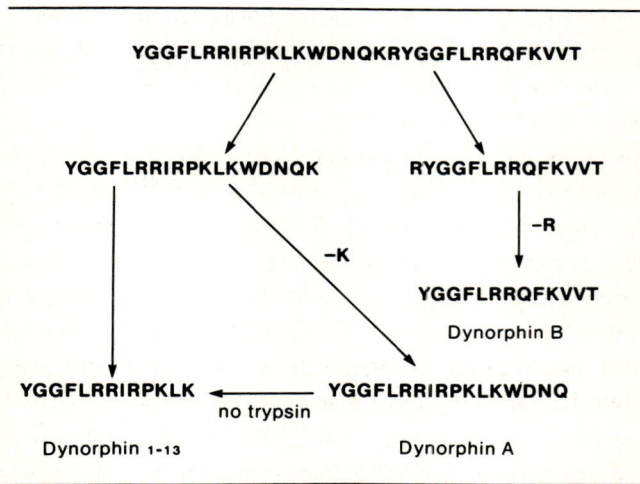

YGGFLRRIRPKLKWDNQKRYGGFLRRQFKVVT

YGGFLRRIRPKLKWDNQK   RYGGFLRRQFKVVT

−R

YGGFLRRQFKVVT
Dynorphin B

−K

YGGFLRRIRPKLK ← YGGFLRRIRPKLKWDNQ
no trypsin

Dynorphin 1-13   Dynorphin A

Fig. 2.5. Dynorphinergic pathways.

The peptide bond between ^{18}Lys and ^{19}Arg in this large dynorphin precursor is significant because trypsin-like cleavage of that bond generates an amino terminal heptadecapeptide (dynorphin A) and a carboxy terminal tridecapeptide (dynorphin B). The latter peptide, dynorphin B, represents the third "big" LE besides dynorphin A and alpha-neoendorphin (Tyr-Gly-Gly-Phe-Leu-Arg-Lys-Tyr-Pro-Lys). The dynorphin B molecule suprisingly displays the same high kappa receptor activity as dynorphin A, a fact which indicates that the "address" needed for receptor recognition and binding most probably resides in the N-termini of the two peptides. The three fragments have been mapped in rat brain with immunocytochemical techniques (73).

As stated above, recent biochemical and anatomical studies provide evidence that the opioid peptidergic systems in the brain are comprised of the enkephalin, dynorphin, and beta-endorphin systems (74). Both the enkephalin and dynorphin peptidergic systems occur in many of the same areas, the enkephalin system but not the dynorphin system occurs in a large number of areas, and a few areas of nuclei contain the dynorphin but not the enkephalin system (75a). This study clarifies the relationship between the enkephalinergic and dynorphinergic neuronal systems.

2.10 OPIOID RECEPTORS

The stereochemical, physical, and ionic aspects of the analgesic receptor surface are outlined (48). Four aspects are considered - anionic site, charge focus, cavity, and flat surface. Figure 2.3 outlines the structural features of the receptor surface.

It is suggested that multiple opiate receptors exist because it is observed that benzomorphan (Figure 2.2) opiates have definite agonist analgesic activity, yet fail to substitute for morphine in suppressing morphine withdrawal systems. The suggestion is made that benzomorphan acts at kappa receptors as opposed to those receptors where morphine is the prototypical agonist-the mu opiate receptor site (75b). Other opiates elicit hallucinogenic effects and it is postulated that a third opiate receptor, sigma, exists. While all three of the opiate receptors may be blocked by naloxone, higher doses of naloxone are required to block the kappa and the sigma receptors, as opposed to the lower amounts needed to block the mu receptor. Table 2.4 collects a list of the various receptors and their agonists and antagonists (26).

Several experimental data indicate that the mu and delta receptors are constructed by physically distinct macromolecules. These opiate receptors

TABLE 2.4.   List of the different classes of opioid receptors and their agonists and antagonists.

RECEPTOR		DRUG
mu$_{1,2}$	agonists	morphine, oxymorphone, fentanyl, dihydromorphine, FK-33, 824
	antagonists	naloxone, levallorphan
kappa	agonists	ethylketazocine, dynorphin
kappa/sigma	agonists	cyclazocine, nalorphine, pentazocine
sigma	agonists	SKF-10, 047
mu/sigma	agonists	etorphine, phenazocine
delta	agonists	met^5-enkephalin, D-ala-2-D-leu^5-enkephalin
epsilon	agonists	beta-endorphin
iota	agonists	enkephalin (intestine)

are regulated by both sodium ion and guanine nucleotides, where both compounds decrease the affinity of the agonists, but not the antagonists, for those binding sites. The opiate agonists which are affected most by sodium correspond to the mu class, while the delta-selective drugs are affected to a lesser extent.

ME and LE are localized in separate neuronal populations in the brain as evidenced by immunohistochemical data. Since these two peptides are localized at different synapses, the question arises whether different methionine enkephalinergic and leucine enkephalinergic receptor sites exist and whether they correspond specifically to the mu versus delta receptors, respectively. Binding site studies have recently revealed that ME is relatively more selective for the mu receptors than LE, while on the other hand, LE displays some preferential selectivity for delta receptors. In guinea pig brain, the following receptor distribution pattern is found: mu (25%); delta (45%); kappa (30%).

2.11 HIGH PERFORMANCE LIQUID CHROMATOGRAPHY OF PEPTIDES

A combination of HPLC and RIA was utilized to determine a number of endorphin-like peptides in rat pituitary gland and the presence of des-tyrosine endorphins was demonstrated for the first time (76). The HPLC of opioid peptides extracted from porcine brain or CSF has been discussed. The lower detection sensitivity of UV absorbance at 280 nm was utilized where detection sensitivity is limited to approximately 500 pmol (37). This result compares to 500 fmol sensitivity observed at 200 nm for a solution of synthetic somatostatin (78).

HPLC is combined with RIA to achieve a high level of sensitivity with improved molecular specificity. This analytical method allows separation and detection of multiple forms of immunoactive peptides. N-acetylated beta-endorphin$_{1-27}$, which is inactive at the opiate receptor, is the predominant form of the beta-endorphin peptide in rat intermediate lobe, with beta-endorphin $_{1-31}$ representing a low proportion of the immunoreactivity (79). This study shows clearly, as dynorphin does, that a free N-terminus is required for receptor addressing and binding, whereas an RIA antibody may be raised to the mid-portion of a peptide.

A significant increase in the molecular specificity of the analytical measurement of endogenous ME was obtained when a combination of both dipeptidylaminopeptidase I (DAPI) enzymolysis and gas chromatographic (GC)-mass fragmentographic assay was used for identification of the two dipeptides and the amino acid which are released from enzymolysis of the opioid pentapeptides (80). Quantification of the two resultant dipeptides (YG and GF) and one amino acid (M) released after DAPI enzymatic digestion of ME was obtained by utilizing the selected ion monitoring (SIM) technique with deuterated methionine as internal standard. Pentafluorinated derivatives were utilized and either a quadrupole or magnetic (LKB 9000) combination GC-MS instrument was used for measurement. Rats were sacrificed by microwave irradiation focused to the cranium. Beef adrenal medulli were obtained from the abattoir and kept frozen until exposed to microwave irradiation. Heat-inactivated DAPI did not release the appropriate dipeptides from an enkephalin. The sensitivity of this method of detection was approximately eight ng for the two dipeptides YG and GF and one ng for the amino acid methionine.

2.12 MODIFIED ENKEPHALINS AND ENDORPHINS

In addition to the enkephalins and endorphins which have unmodified N- and C-termini and amino acid residues, post-translational modification of these gene products also indicates that a variety of chemical groups is available to modify various portions of the peptide molecule. These modification sites include the N-terminus, C-terminus, the tyrosine residue, the thioether sulfur atom of methionine, among others. N-acetyl LE accounts for 20-25% of the total LE immunoreactivity in the rat neurointermediate pituitary. This N-acetyl LE is opiate inactive and indicates that N-acetylation in the rat pituitary is not restricted to only beta-endorphin, but rather may be a general metabolic control mechanism. Other areas (hypothalamus, striatum, midbrain, and pons-medulla) of the rat brain did not contain any N-acetylated derivative (81).

Both ME-like and LE-like immunoreactivities were detected at an early fetal age in both brain and gut tissues of the rat (82). The authors speculate that this immunoreactivity derives from a Tyr-O-sulfate derivative of the enkephalins. At such an early developmental age, apparently more LE than ME is present, whereas the reverse situation is commonly true in the adult brain. The Tyr-O-sulfate residue was found to be hydrolytically sensitive to the trifluoroacetic acid used in the extraction. Both protein-incorporated LE sulfate and the free pentapeptide were found in sheep, mouse, and guinea pig striatum, where approximately 10-20% of the total free pool of LE is sulfated. The corresponding sulfate ester of ME has not been discovered by this methodology. Sulfation may represent either an inactivation mechanism, a signal for selective precursor cleavage, or even a means to confer novel non-opiate or other opiate activities on these opiate peptides (83).

The thioether group in ME is capable of spontaneous oxidation to the sulfoxide (84). The peptidase-resistant ME analog, D-ala^2-met^5-enkephalin amide, which is a pharmacologically active analog due to its prolonged duration of action, may also be spontaneously oxidized. Indeed, the opioid activity of the analog is clearly enhanced by oxidation of the thioether in the terminal methionine residue. These results demonstrate the crucial experimental fact that one should always provide chromatographically-pure compounds for any pharmacological studies, because an equieffective dose of closely-related compounds may have drastically different activities. Indeed, as this experimentally significant theme develops through the remainder of this book, it will be amply demonstrated that MS should be used to at least confirm an appropriate molecular weight for the peptide, if indeed it is not also used to confirm the amino acid sequence. The sulfoxide derivative probably does not possess a higher affinity for the opiate receptor, but rather may have a greater resistance to peptidase-degrading activity.

The finding that ME sulfoxide has greater biologic activity compared to the thioether enkephalin indicates the possibility that another, higher order of elegant metabolic control is available to a tissue and involves an enzymatic step to reduce a methionine sulfoxide to a methionine amino acid residue. An assay has been developed for peptide methionine sulfoxide reductase activity (85) and it is conceivable, but not yet demonstrated experimentally, that this sulfoxide reductase system participates in a "fine-tuning" metabolic control available to a tissue and involves the activation and inactivation of a ME thioether - sulfoxide redox system.

2.13 URINARY PEPTIDES

Urinary peptide metabolic profile patterns have been determined for a variety of psychiatric disorders (depression, schizophrenia, autism, hyperkinesia, stress, anorexia nervosa; 86). Patients suffering from anorexia nervosa apparently involve a heterogenous population, and five different chromatographic patterns were found corresponding to five different behavioral patterns (86). The physiological hypothesis stated by these authors indicates that either formation or release of peptides may surpass genetically-determined metabolism capacity of those patients and that the overflow of these peptides to body fluids and urine may be noted. A review of the literature relating to psychiatric research indicates that, on one hand, an excess of beta-endorphin may be responsible for the occurrence of catatonia whereas a critical balance amongst alpha-, beta-, and gamma-endorphins is physiologically important wherein an imbalance may result in schizophrenia (87). The neurochemical pathology of brain peptides has been described while focusing on the complex number of routes to post-translational modifications (88).

2.14 CEREBROSPINAL FLUID PEPTIDES

A number [32] of peptides which are putative neurotransmitter/ neuromodulators in CSF have been measured in humans (89a). Table 2.5 lists those peptides which are tentatively identified and measured in human CSF.

CSF is hypothesized to serve not only as a waste disposal system for neuropeptides, but also as a more active system of biochemical and behavior relevance. It is noted that several circumventricular structures and organs apparently are highly specialized for transportation of substances between brain and CSF as well as between CSF and blood. The caudate nucleus represents one structure which is in intimate contact with the third ventricle. Larger prohormones and smaller metabolic products are measurable in cerebrospinal fluid. CSF also plays roles as a mechanical cushion, transport, and waste disposal system.

The opioid activity in CSF samples obtained from patients undergoing diagnosis for intracranial hydrodynamic dysfunction were fractionated (Sephadex G10) and electrophoretically separated. The characteristics of endorphins identified by RIA and RRA were compared (89b). While RIA did not detect any of the opioid peptides, RRA detected a complex pattern of opioid receptor-active peptides, some of which seem to be metabolically related to each other. The hypothesis is that peptides in one area (FII) of the chromatogram derive from an enkephalin precursor while those from

TABLE 2.5. Peptides tentatively identified and measured in human CSF (89a).

1. 16 K fragment	17. substance P
2. beta-lipotropin	18. neurotensin
3. adrenocorticotrophic hormone (ACTH)	19. somatostatin (SRIF)
4. beta-endorphin	20. gastrin
5. leucine-enkephalin	21. cholecystokinin (CCK)
6. methionine-enkephalin	22. vasoactive intestinal peptide
7. alpha-melanocyte stimulating hormone	23. bradykinin
8. beta-melanocyte stimulating hormone	24. bombesin
9. calcitonin	25. thyrotropin-releasing hormone (TRH)
10. oxytocin	26. thyroid-stimulating hormone (TSH)
11. vasopressin (VP)	27. luteinizing-hormone releasing hormone (LHRH)
12. neurophysin	28. luteinizing hormone (LH)
13. arginine vasotocin (AVT)	29. follicle-stimulating hormone (FSH)
14. angiotensin I	30. growth hormone (GH)
15. converting enzyme	31. prolactin
16. angiotensin II	32. insulin

another area (FI) may derive from the dynorphin system. It was noted that there is apparently a complex and critical relationship amongst several experimental parameters - anatomically defining neuronal distribution of opioid peptides, CSF sampling procedures, technical constraints associated with that sampling, and the functional state in the neurons being investigated (See also Chapter 3).

Data are presented to support the positive correlation between endorphin levels in CSF and measures of pain (90). Patients with chronic pain syndromes were studied for endorphin levels, CSF, and pain measurements (electro-stimulation via saline electrodes). In patients with a high level of fraction I (FI), pain threshold and tolerance were significantly higher than in patients with low levels of that chromatographic fraction. Pain relief may be achieved by acupuncture (91), counterirritation, or electrical stimulation, where those responses suggest that pain relief may be due to activation of endogenous antinociceptive processes, presumably consisting of the triad of enkephalinergic, endorphinergic, and dynorphinergic systems in an effort to decrease the influence of the noxious stimulus. One hypothesis is that opioid peptide-containing neurons for neuromodulatory activity synapse with, and also decrease the firing rate of, substance P-containing neurons (92-94). Indeed, if the action of the analgesia-producing system is mediated

by endogenous morphine-like compounds such as the enkephalins and
endorphins, then a disruption of this peptidergic system by the pure opiate
antagonist naloxone (Figure 2.2) would increase the perceived intensity of
the maintained stimulant (95). Experimental evidence supports the hypothesis
that pain is an important activating factor of the endorphin-mediated
analgesia system to, in turn, decrease the effects of the noxious stimulus.

An interesting clinical situation may impact on and provide beneficial
results to studies of endorphins. A family (96) and one patient (97) have
been found with the congenital absence of pain. Two types of congenital
analgesia may be described, either insensitivity to or indifference to pain.
Insensitivity to pain is the result of abnormalities of peripheral nerves or
central sensory pathways, while patients who are indifferent to pain have
normal sensory paths in peripheral nerves but fail to appreciate pain stimuli
where the primary lesion involves central structures where pain is integrated.
In these patients, endorphin levels RIA in serum and CSF were within normal
limits whereas CSF total opioid activity RRA was elevated.

2.15 PEPTIDES IN NUTRITION

Experimental evidence from animal studies indicates that opioid peptides
may be involved in nutrition of the body via a regulatory process involving
appetite and/or satiety (98). Table 2.6 lists the peptides found in the CNS.

2.16 SYNTHESIS OF PEPTIDES

While peptide synthesis is an entire field of research unto itself and
will not be discussed here, brief mention will be made of the fact that solid
phase techniques are utilized in a study to synthesize novel analogs of the
enkephalins and to identify those functional groups in a peptide which are
required for biologic activity. This type of study derives from the fact
that many conformational studies have proposed receptor-bound conformations
for enkephalins, whereby the mammalian- derived pentapeptides (Table 2.1) and
the plant-derived opioid alkaloids (Figure 2.2) both pharmacologically compete
for the same in vivo receptors (99). This type of structural study attempts to
take advantage of the presumed electrostatic interactions which are the basis
of structure-function correlation with the three-dimensional polypeptide
conformation of the receptor site (100).

2.17 OTHER NEUROPEPTIDES

In addition to the larger amount of research activity recently invested
in the endorphin, enkephalin, and dynorphin systems, the structures of other

TABLE 2.6. Peptides found in the CNS (98).

HYPOPHYSIOTROPIC HORMONES
   1. thyrotropin-releasing hormone (TRH)
   2. luteinizing hormone/follicle-stimulating hormone releasing
      hormone, or gonadotropin releasing hormone (GnRH)
   3. growth hormone release-inhibiting hormone (somatostatin)

ADENOHYPOPHYSIAL HORMONES (including opiocortins)
   1. growth hormone
   2. prolactin
   3. thyroid stimulating hormone (TSH)
   4. luteinizing hormone
   5. follicle-stimulating hormone
   6. opiocortins
     a. adrenocorticotropic hormone (ACTH)
     b. beta-lipotropic hormone (beta-LPH)
     c. endorphins
     d. enkephalins
     e. sigma-melanocyte stimulating hormone (sigma-MSH)
     f. beta-melanocyte stimulating hormone (beta-MSH)

NEUROHYPOPHYSIAL HORMONES
   1. vasopressin or antidiuretic hormone (ADH)
   2. oxytocin
   3. neurophysins

BRAIN-GUT HORMONES
   1. cholecystokinin (CCK)
   2. gastrin
   3. insulin
   4. vasoactive intestinal polypeptide (VIP)
   5. motilin
   6. bombesin
   7. substance P (SP)
   8. neurotensin
   9. glucagon

OTHERS
   1. bradykinin
   2. isorenin-angiotensin
   3. sleep peptide
   4. carnosin
   5. calcitonin

---

neuropeptides have also been elucidated. These neuropeptides include somatostatin (101), cholecystokinin (102), neurotensin (103), corticotropin-releasing factor (CRF) (104, 105), substance P, delta sleep-inducing peptide (106), and others.

It is apparent from this partial list that, when listing biologically active peptides, it is a common practice to name that peptide by the first perceived biologic function that it exhibits. Researchers are becoming more aware that this procedure of naming peptides by only one biologic function may inhibit further research into other possible bioactivities that that peptide may also possess. While it is convenient to name peptides by their biological effects, and in all probability researchers will continue to do so, at least researchers must not be circumscribed in their thinking of other pertinent experiments by the name of a specific peptide (107).

2.18 CONCENTRATIONS OF ENDOGENOUS PEPTIDES

In general, peptide hormones are present in the brain at the lower concentration level of pmol peptide mg^{-1} protein compared to classical biogenic amine neurotransmitters (nmol mg^{-1}) and amino acid neurotransmitters (mmol mg^{-1} protein) (108,109). It is important to realize that the CNS concentration of peptides which are measured in homogenized tissues may not reliably indicate either the microenvironmental functional importance or the turnover characteristics of those peptides because regional, cellular, and subcellular distributions of peptides are probably of more in vivo biological relevance. While some researchers believe that disparate tissues such as brain and gut derive from the neural crest and that this is the developmental reason why similar peptides are found at these two distant locations, other workers have shown that this rationalization of similarity of peptide genesis may not be the case. There are several modes possible for peptide secretion in the gastrointestinal (GI) tract, pancreas, pituitary, and neural tissue. As mentioned above, neurotransmitters or neuromodulators may occur in the neuronal system. However, if the neuronal product is released into the blood stream, then it is a neurohormone.

A study involving regional concentrations of selected peptides indicates noticeable and, in some cases, marked concentration differences. For example, somatostatin ranges from (in pmol g^{-1} wet weight tissue) 1,000 in the hypothalamus, to 23 in the substantia nigra, and to 1 in the cerebral cortex, whereas substance P has a maximum concentration of 922 in the substantia nigra, 518 in the globus pallidus, and only 0.3 in the cerebellar cortex. These values are distinguished from those of neurotensin which optimize in the hypothalamus median eminence (43 pmol), through the substantia nigra (23 pmol), down to the cerebella cortex (0.8) (108,109). One can readily infer from these data that the regional distribution of these peptides is required so that a microenvironment is provided for the peptides to act in an appropriate neurochemical fashion (with a receptor, precursor formation or metabolism, packaging for transport, etc.).

2.19 SUMMARY

This chapter surveys only a few selected facets of biologically active brain peptides (110). This overall survey reflects an explosive growth in our knowledge of the three peptidergic pathways, the variety of their corresponding molecular precursors, and the number of metabolic products of these biologic compounds. The pathways for opiate peptides now centers on four different peptidergic pathways in the brain: endorphinergic, dynorphinergic, enkephalinergic, and substance P-ergic. The first three peptide pathways are available to a cell to effectively and rapidly deal with noxious stimuli and to interact with substance P-containing neurons. The four peptides (beta-endorphin, dynorphin, enkephalins, substance P) derive from larger precursors where they are embedded in those large precursors. The smaller peptides are bracketed by pairs of basic amino acids, specific enzymes are biologically available to excise individual peptides which are then used to satisfy specific and perhaps variable metabolic demands, and a diversity of post-translational modifications which are specific to particular amino acid residues is available. While this survey may seem relatively complicated, an appreciation is demanded of the distinct specific and elegant metabolic control points in a cell and also indicates that the organism has at its disposal numerous metabolic fine-tuning control mechanisms to deal with the many aspects of pain.

REFERENCES

1   B. Weinstein and S. Lande (Editors), Peptides: Chemistry and
     Biochemistry, Marcel Dekker, Inc., New York, 1970, 538 pp.
2   S. Lande (Editor), Progess in Peptide Research, Vol. II, Gordon
     and Breach, New York, 1972, 393 pp.
3   J. Meienhofer (Editor), Chemistry and Biology of Peptides: Proc.
     Third Amer. Pept. Sympos., Ann Arbor Science, Ann Arbor,
     Michigan, 1972, 762 pp.
4   M. Goodman and J. Meienhofer (Editors), Peptides: Proc. Fifth
     Amer. Pept. Sympos., Halsted Press, New York, 1977, 612 pp.
5   E. Gross and J. Meienhofer (Editors), Peptides: Structure and
     Biological Function, Proc. Sixth Amer. Pept. Sympo., Pierce
     Chemical Company, Rockford, Illinois, 1979, 1079 pp.
6   D.H. Rich and E. Gross (Editors), Peptides:
     Synthesis-Structure-Function, Pierce Chemical Company, Rockford,
     Illinois, 1981, 853 pp.
7   E.R. Kandel and J.H. Schwartz (Editors), Principles of Neural
     Science, Elsevier/North-Holland, New York, 1981, 731 pp.
8   G.J. Siegel, R.W. Albers, B.W. Agranoff and R. Katzman (Editors),
     Basic Neurochemistry, Third Edition, Little, Brown and Company,
     Boston, (1972) 857 pp.
9   D.T. Krieger and J.C. Hughes (Editors), Neuroendocrinology, Sinauer
     Associates, Inc., Sunderland, Massachusetts, 1980, 352 pp.
10   D.T. Krieger, M.J. Brownstein, and J.B. Martin (Editors), Brain
     Peptides, Wiley, N.Y., 1983, 1032 pp.
11   M.O. Dayhoff (Editor), Atlas of Protein Sequence and Structure 1972,
     Vol. 5, National Biomedical Research Foundation, Washington, D.C.,
     1972, 124 pp (plus 418 pp data section).

12   J.B. Martin, S. Reichlin and K.L. Bick (Editors), Neurosecretion and
      Brain Peptides 28, Raven Press, New York, 1981.
13   E. Costa and M. Trabucchi (Editors), Neural Peptides and Neuronal
      Communication, Adv. Biochem. Psychopharmacol., Vol. 22,
      Raven Press, N.Y., 1980, 651 pp.
14   E. Costa and M. Trabucchi (Editors), Regulatory Peptides-From
      Molecular Biology to Function, Raven Press, N.Y. 1982.
15   C.A. Marsan and W.Z. Traczyk (Editors), Neuropeptides and Neural
      Transmission, Vol. 7, Raven Press, New York, 1980, 391 pp.
16   H.H. Loh and D.H. Ross, (Editors), Neurochemical Mechanisms of
      Opiates and Endorphins, Vol. 20, Raven Press, New York 1979, 563 pp.
17   R. Porter and M. O'Conner (Editors), Substance P in the Nervous
      System, Pitman Press, London, 1982, 349 pp.
18   R.F. Beers and E.G. Bassett (Editors), Polypeptide Hormones, Raven
      Press, New York, 1980, 528 pp.
19   E.M. Rodriguez and Tj.B. van Wimersma Greidanus (Editors),
      Cerebrospinal Fluid (CSF) and Peptide Hormones, S. Karger, Basel,
      1982, 220 pp.
20   G. Ungar (Editor), Molecular Mechanisms in Memory and Learning,
      Plenum Press, New York, 1970, 296 pp.
21   F.E. Bloom (Editor), Peptides: Integrators of Cell and Tissue
      Function, Raven Press, New York, 1980, 257 pp.
22   J.C. Liebeskind, R.K. Dismukes, J.L. Barker, P.A. Berger, I.
      Creese, A.J. Dunn, D.S. Segal, L. Stein and W.W. Vale (Editors),
      Neurosci. Res. Prog.Bull. 16, December, 1978, MIT Press, pp. 490-634.
23   H.W. Kosterlitz and L.Y. Terenius (Editors), Pain and Society, Verlag
      Chemie, Weinheim, 1980.
24   A.M. Gotto, E.J. Peck and A.E. Boyd (Editors), Brain Peptides: A
      New Endocrinology, Elsevier/North Holland Biomedical Press,
      Amsterdam, 1979, 406 pp.
25   L.L. Iversen, R.A. Nicoll and W.W. Vale, Neurosci. Res. Prog.
      Bull. 16, June, 1978, MIT Press, Cambridge, MA, pp. 211-370.
26   G.A. Olson, R.D. Olson, A.J. Kastin and D.H. Coy, Peptides, 3
      (1982) 1039-1072.
27   R.C.A. Frederickson and L.E. Geary, Cong. Neurobiol., 19 (1982)
      19-69.
28   H. Akil and S.J. Watson, in Handbook of Psychopharmacology, Vol.
      16, L. Iverson, S. Iverson and S.H. Snyder (Editors), Plenum, 1983.
29   R.J. Miller, in L. Iversen, S.D. Iversen and S.H. Snyder
      (Editors), Handbook of Psychopharmacology, Vol. 16, Plenum ,1983.
30   S.E. Leeman, N. Aronin and C. Ferris, Rec. Progr. Horm. Res.,
      38 (1982), 93-132.
31   W.M. Cowan, Z.W. Hall and E.R. Kandel (Editors), Ann. Rev.
      Neurosci. 2, Annual Reviews, Inc., Palo Alto, California, 1979,
      555 pp.
32   W. Voelter and G. Weitzel (Editors), Structure and Activity of Natural
      Peptides, W. de Gruyter, Berlin, 1981, 634 pp.
33   A. Eberle, R. Geiger and T. Wieland (Editors), Perspectives in
      Peptide Chemistry, S. Karger, Basel, 1981, 444 pp.
34   P.G. Katsoyannis (Editor), The Chemistry of Polypeptides,
      Plenum Press, New York, 1973, 417 pp.
35   H. Siegel and R.B. Martin, Chem. Rev., 82 (1982) 385-426.
36   L. Heimer, The Human Brain and Spinal Cord, Springer-Verlag, New
      York, 1983, pp. 327-328.
37   W.J.H. Nauta and M. Freitag, in The Brain, Freeman, San Francisco,
      1979, 55 pp.
38   F.H. Netter, in Nervous System, CIBA, Case-Hoyt, Rochester, NY,
      1977, 168 pp.

40

39   R. Burgus, T. Dunn, D.M. Desiderio, W. Vale and R. Guillemin,
      Compt. Rend., 269 (1969) 226-228.
40   R. Burgus, T. Dunn, D.M. Desiderio and R. Guillemin, Compt.
      Rend., 269 (1969) 1870-1873.
41   R. Burgus, T. Dunn, D.M. Desiderio, D.N. Ward, W. Vale, R.
      Guillemin, A.M. Felix, D. Gillessen and R. Studer, Endocrinol., 86
      (1970) 573-582.
42   R. Burgus, T.F. Dunn, D.M. Desiderio, D.N. Ward, W. Vale, and R.
      Guillemin, Nature, 226 (1970) 321-325.
43   D.M. Desiderio, R. Burgus, T.F. Dunn, D.N. Ward, W. Vale and R.
      Guillemin, Org. Mass Spectrom., 5 (1971) 221-228.
44   J. Hughes, T.W. Smith, H.W. Kosterlitz, L.A. Fothergill, B.A.
      Morgan and H.R. Morris, Nature, 258 (1975) 577-579.
45   M. Rubinstein, S. Stein and S. Udenfriend, Proc. Natl. Acad. Sci.
      USA, 74 (1977) 4969-4972.
46   S. Zakarian and D.G. Smyth, Nature, 296 (1982) 250-252.
47   S.H. Snyder, Chem & Eng. News, Nov. 28, 1977, 26-35.
48   A.H. Beckett, Fortschr. Arzneimit., 1 (1959) 455-530.
49   J.D. Barchas, H. Akil, G.R. Elliott, R.B. Holman and S.J. Watson,
      Science, 200 (1978) 964-973.
50   T. O'Donohue and D.M. Dorsa, Peptides, 3 (1982) 353-395.
51   S.F. Atweh, in Smith and Lane (Editors), The Neurobiology of Opiate
      Reward Processes, Elsevier Biomed. Press, Amsterdam, 1983, pp. 59-88
52   D.M. Desiderio, H. Onishi, H. Takeshita, F. S. Tanzer, C. Wakelyn,
      J.A. Walker, Jr. and G. Fridland, J. Neurochem., submitted.
53   J. Rossier, A. Bayon, T.M. Vargo, N. Ling, R. Guillemin and F.
      Bloom, Life Sci., 21 (1977) 847-852.
54   D. Wesche, V. Hollt and A. Herz, Arch. Pharmacol., 301 (1977) 79-82.
55   C. Gros, P. Pradelles, C. Rouget, O. Bepoldin, F. Dray, M.C.
      Fournie-Zaluski, B.P. Roques, H. Pollard, C. Llorens-Cortes and J.C.
      Schwartz, J. Neurochem., 31 (1978) 29-39.
56   V. Clement-Jones, P.J. Lowry, L.H. Rees and G.M. Besser, J.
      Endocrinol., 86 (1980) 231-243.
57   A. Dupont, J. Lepine, P. Langelier, Y. Merand, D. Rouleau, H.
      Vaudry, C. Gros and N. Barden, Reg. Peptides, 1 (1980) 43-52.
58   M.J. Kubek and J.F. Wilber, Neurosci. Lett., 18 (1980) 155-161.
59   P.C. Emson, A. Arregui, V. Clement-Jones, B.E.B. Sandberg and M.
      Rossor, Brain Res., 199 (1980) 147-160.
60   M.S.A. Kumar, C.L. Chen and H.H. Huang, Neurobio. Aging, 1 (1980)
      153-155.
61   E. Weber, R.J.W. Truscott, C. Evans, S. Sullivan, P. Angwin and
      J.D. Barchas, J. Neurochem., 36 (1981) 1977-1985.
62   I. Lindberg and J.L. Dahl, J. Neurochem., 36 (1981) 506-512.
63   A.S. Liotta, J. Yamaguchi and D.T. Krieger, J. Neurosci., 1 (1981)
      585-595.
64   H. Akil, E. Young, S.J. Watson and D.H. Coy, Peptides, 2 (1981)
      289-292.
65   S.P. Kalra, Neuroendocrinology, 35 (1982) 28-31.
66   A.M. Di Giulio, G.B. Picotti, A.M. Cesura, A.E. Panerai and P.
      Mantegassa, Life Sci., 30 (1982) 1605-1614.
67   B.J. Davis, G.D. Burd and F. Macrides, J. Comp. Neurol., 204
      (1982) 377-383.
68   A. Goldstein, S. Tachibana, L.I. Lowney, M. Hunkapiller and L.
      Hood, Proc. Natl. Acad. Sci. USA, 76 (1979) 6666-6670.
69   B.R. Seizinger, V. Hollt and A. Herz, Biochem. Biophys.
      Res.Commun., 102 (1981) 197-205.
70   A. Goldstein, W. Fischli, L.I. Lowney, M. Hunkapiller and L. Hood,
      Proc. Natl. Acad. Sci. USA, 78 (1981) 7219-7223.

71 C. Chavkin, I.F. James and A. Goldstein, Science, 215 (1982) 413-415.
72 W. Fischli, A. Goldstein, M.W. Hunkapiller and L.E. Hood, Proc. Natl. Acad. Sci. USA, 79 (1982) 5435-5437.
73 S.J. Watson, H. Khachaturian, L. Taylor, W. Fischli and H. Akil, Proc. Natl. Acad. Sci, 80 (1983) 891-894.
74 S.J. Watson, H. Khachaturian, H. Akil, D.H. Coy and A. Goldstein, Science, 218 (1982) 1134-1136.
75a S.R. Vincent, T. Hokfelt, I. Christensson and L. Terenius, Neurosci. Lett., 33 (1982) 185-190.
75b S.H. Snyder and R.R. Goodman, J. Neurochem., 35 (1980) 5-15.
76 J.G. Loeber, J. Verhoef, J.P.H. Burbach and A. Witter, Biochem. Biophys. Res. Commun., 86 (1979) 1288-1295.
77 A. Dell, T. Etienne, H.R. Morris, A. Beaumont, R. Burrell and J. Hughes, in MacIntyre and Szelke, (Editors), Molecular Endocrinology, Elsevier, North-Holland, Biomedical Press, 1979. pp. 91-97.
78 D.M. Desiderio and M.D. Cunningham, J. Liq. Chromatogr., 4 (1981) 721-722.
79 H. Akil, Y. Ueda, H.L. Lin and S.J. Watson, Neuropeptides, 1 (1981) 429-446.
80 E. Peralta, H.-Y. Yang, J. Hong and E. Costa, J. Chromatogr., 190 (1980) 43-51.
81 B.R. Seizinger, V. Hollt and A. Herz, Biochem. Biophys. Res. Commun., 101 (1981) 289-297.
82 J.L. Dahl, M.L. Epstein, B. Silva and I. Lindberg, Life Sci., 31 (1982) 1853-1856.
83 C.D. Unsworth and J. Hughes, Nature, 295 (1982) 519-522.
84 J.A. Kiritsy-Roy, S.K. Chan and E.T. Iwamoto, Life Sci., 32 (1983) 889-893.
85 N. Brot, J. Werth, D. Koster and H. Weissbach, Anal. Biochem., 122 (1982) 291-294.
86 O.E. Trygstad, K.L. Reichelt, I. Foss, P.D. Edminson, G. Saelid, J. Bremer, K. Hole, H. Orbeck, J.H. Johansen, J.B. Boler, K. Titlestad and P.K. Opstad, Brit. J. Psychiat., 136 (1980) 59-72.
87 J.M. van Ree and D. De Wied, Neuropharm., 20 (1981) 1271-1277.
88 J.A. Edwardson and J.R. McDermott, Brit. Med. Bull., 38 (1982) 259-264.
89a F. Nyberg and L. Terenius, Life Sci., 31 (1982) 1740.
89b R.M. Post, P. Gold, D.R. Rubinow, J.C. Ballenger, W.E. Bunney and F.K. Goodwin, Life Sci., 31 (1982) 1-15.
90 L.V. Knorring, B.G.L. Almay, F. Johansson and L. Terenius, Pain, 5 (1978) 359-365.
91 J.S. Han and L. Terenius, Ann. Rev. Pharmacol. Toxicol., 22 (1982) 193-220.
92 N. Aronin, M. Difiglia, and S.E. Leeman, in D.T. Krieger, M.J. Browstein, and J.B. Martin (Editors), Brain Peptides, Wiley, N.Y., 1032.
93 R.A. Nicoll, C. Schenker and S.W. Leeman, Ann. Rev. Neurosci., 3 (1980) 268 pp.
94 J.L. Henry, B.J. Sessle, G.E. Lucier, and J.W. Hu, Pain 8 (1980) 33-45
95 J.D. Levine, N.C. Gordon, R.T. Jones and H.L. Fields, Nature, 272 (1978) 826-827.
96 M.A. Fath, M.R. Hassanein and J.I.P. James, J. Bone Joint Surg., 65B (1983) 186-188.
97 M. Manfredi, G. Bini, G. Cruccu, N. Accornero, A. Berardelli and L. Medolago, Arch. Neurol., 38 (1981) 507-511.
98 A. Kitabchi, in Hormone Secreting Pituitary Tumors, James R. Givens, Jr. (Editor), Yearbook Medical Publishers, Chicago, 1982, pp 45-62.

99   F.A. Gorin, T.M. Balasubramanian, T.J. Cicero, J. Schwietzer and
     G.R. Marshall, J. Med. Chem., 23 (1980) 1113-1122.
100 A. Warshel, Acc. Chem. Res., 14 (1981) 284-290.
101 J. Spiess, J.E. Rivier, J.A. Rodkey, C.D. Bennett and W. Vale,
     Proc. Natl Acad. Sci., 76 (1979) 2974-2978.
102 J.E. Morley, Life Sci., 30 (1982) 479-493.
103 G. Bissett, P. Manberg, C.B. Nemeroff and A.J. Prange, Life Sci.,
     23 (1978) 2173-2182.
104 W. Vale, J. Spiess, C. Rivier and J. Rivier, Science, 213 (1981)
     1394-1397.
105 A.V. Schally, R.C.C. Chang, A. Arimura, R.W. Redding, J.B.
     Fishback and S. Vigh, Proc. Natl. Acad. Sci. USA, 78 (1981)
     5197-5201.
106 W.A. Banks, A. Kastin and D.H. Coy, Pharmacol. Biochem. Behav.,
     17 (1982) 1009-1014.
107 A.J. Kastin, W.A. Banks, J.E. Zadina and M. Graf, Life Sci., 32
     (1983) 295-301.
108 D.T. Krieger and J.B. Martin, New Eng. J. Med., 304 (1981) 876-885.
109 D.T. Krieger and J.B. Martin, New Eng. J. Med., 304 (1981) 944-951.
110 D.T. Krieger, Science, 222, (1983) 975-985.

Chapter 3
BIOCHEMICAL SAMPLING TECHNIQUES

3.1 INTRODUCTION

This chapter will discuss several experimental factors that must be considered before, during, and after an analytical measurement is taken of any biological compound. These factors are time, post-translational peptide modifications, animal size, species differences, method of sacrifice, internal brain structures, and peptidase activity. Other factors necessary to a particular study must be considered, whenever appropriate.

Additionally, when using an animal model during elucidation of the molecular processes involved in a selected metabolic mechanism, it is prudent to be aware of and take into consideration several ethical and experimental factors. First, the welfare of the animal which is involved in the experiment must be the foremost consideration of the researcher. Appropriate published guidelines must be adhered to, and in some journals (1), experimental data will not be published without certification to that adherence.

3.2 TIME FACTORS

Because of the extremely rapid (msec) timecourse of the biochemical events involved in, for example, pain transmission, it is crucial to be aware of potential and real time-dependent factors which can occur in any of the steps between the intact animal model, tissue procurement, compound purification, and analytical measurement. The most significant temporal factors include enzymolysis/hydrolysis, thioether oxidation-reduction, and other structural modifications (amidation, acetylation, phosphorylation, sulfation, etc.).

Exsanguination is an appropriate experimental method to rapidly inactivate blood enzymes from a larger animal model. Exsanguination can be effected via femoral artery drainage or aortic incision. When sensitivity of the analytical measurement method is increased to a sufficiently high level, a smaller animal is utilized and microwave irradiation may be appropriate for rapid sacrifice and tissue treatment.

After exsanguination, decapitation, or microwave treatment, it is important to extract rapidly the target tissue or fluid. Target tissues and fluids must be cooled (liquid nitrogen, dry ice-acetone) to minimize any rapid biochemical and enzymatic interconversions. Experience with endogenous prostaglandins (2-4) indicates that the crucial aspect of rapid temperature lowering is required to ensure the fact that only endogenous compounds remain present in this critical first step. Enzymatic interconversions of peptides can also occur by unintentionally or unknowingly mobilizing the nociceptive process itself, an event which precipitates a cascade of biochemical events. The first step in production of opioid peptides involves, in one particular pathway, a preproenkephalin molecule comprised of 264 amino acids and which contains six copies of ME one copy of LE (Figure 3.1).

These enkephalin pentapeptides are bracketed in the precursor molecule by pairs of the basic amino acids, Lys (K) and Arg (R), which provide target sites for rapid metabolic sensitivity towards trypsin-like and carboxypeptidase B-like enzymatic activity. Once enzymic excision of the selected peptide occurs, the enkephalin pentapeptides become sensitive to further action of enkephalinases, aminopeptidases, dipeptidylaminopep-tidases, and perhaps other yet-to-be-described peptidases. Exsanguination removes the blood enzymes, and if the tissue is not damaged, tissue proteases will not be released to initiate proteolysis.

It cannot be overemphasized that great care must be taken in this type of experiment while deriving the biologic tissue which will be used for time-consuming and expensive analytical measurements upon which physiological and clinical conclusions will be based and which will be the source for scientific verification in many other laboratories around the world for years to come. This is an important consideration because it is the author's observation that, in general, the following type of situation for

FLGLCTWLLALGPGLLATVRAECSQDC-
SYRLARPTDLNPLACTLESEGKLPSLK-
TCKELLQLTKLELPPDATSALSKQEES-
AKK<u>YGGFM</u>KR<u>YGGFM</u>KKMDELYPLEVE-
NGGEVLGKR<u>YGGFM</u>KKDAEEDDGLGNS-
LKELLGALDQREGSLHQEGSDAEDVSK-
<u>GFMRGL</u>KRSPHLEDETKELQKR<u>YGGFM</u>-
GRPEWWMDYQKR<u>YGGFL</u>KRFAEPLPSE-
ESYSKEVPEMEKR<u>YGGFMRF</u>

Fig.3.1. Amino acid sequence of bovine adrenal preproenkephalin.

collaboration prevails. A "biologically-oriented" researcher does all of the preliminary work and then turns over the sample to a researcher who generally provides chromatographic and analytical experiments. While this collaboration may work well in some cases, great care must be given to the respective needs and understanding of the collaborators.

3.3 POST-TRANSLATIONAL MODIFICATIONS TO PEPTIDES

Another biological event that can structurally alter a target peptide is chemical, biochemical, or enzymatic post-translational modifications of specific amino acid residues. This combination of both chemical and enzymatic interconversions for the structural modification of biological peptides demonstrates the presence of an elegant series of effective metabolic control points which are available to an animal to "fine-tune" its opioid peptide metabolic apparatus in an effort to maintain efficient homeostatic control, or to rapidly mobilize a potential reservoir of peptide. For example, in addition to enzymolysis mentioned above for enkephalins, N-terminal tyrosine residues may be sulfated, amidation of the C-terminus, D, or E residues may occur, and the amino terminus may be acetylated. It was stated above that N-acetylated beta-endorphin$_{1-27}$ is the predominant immunoreactive species in brain tissue even though it has no opioid reactivity because it does not interact with the appropriate opiate receptor (5). This molecule illustrates the type of severe experimental limitation which may be experienced when only an immunoassay is used to measure a biologically important molecule. Lastly, the methylthioether group of a methionine residue is sensitive to oxidation and may readily convert to a methionine sulfoxide group during extraction procedures. The more hydrophilic sulfoxide compound elutes at a lower retention time under RP-HPLC conditions. A known methionine sulfoxide reductase system also exerts metabolic control (6).

This brief discussion and the presentation of selected examples illustrate the presence of an elegant and sensitive fine-tuning metabolic apparatus available in a tissue for rapid packaging, synthesis, and cleavage to produce the "working opioid peptide" which is then rapidly metabolized to inactive metabolites. This latter point is important for the neurotransmitter roles (Table 2.3) that the endogenous opioid peptides may play.

3.4 ANIMAL SIZE

During the development of a particular new analytical measurement method, it is important to consider the experimental parameter of animal size, because the neuroanatomical tissue to be utilized first for development of a procedure and then for analytical measurement of an endogenous compound must provide a sufficient amount of tissue. For example, if one gram of caudate nucleus tissue is required for development of an MS assay method, then the canine animal model can be conveniently utilized, because the rodents provide too little tissue, whereas the ovine or porcine models would provide an excess. For development of an RRA using limbic system synaptosomes, the canine model again is appropriate, because one animal provides 15g of tissue. The normal variety of readily accessible animal models includes the rodent, canine, ovine, and porcine species.

3.5 SPECIES DIFFERENCES

Whenever choosing an animal model for development of an analytical measurement method, or when measuring a particular target compound, it is significant to remember that species differences may exist in peptide structures (7). These species differences are also manifested in neuroanatomical relationships. For example, the pituitary in the human versus other animals is different, and important neurochemical differences also exist. For example, depending on conservation of amino acid sequences across different species, certain amino acid residues may be replaced by others (see Chapter 2). These amino acid sequence alterations may or may not play a significant role when a particular assay method is discussed. For example, it may be possible that an RIA antibody may be sensitive to that portion of the molecule which underwent an amino acid replacement. The BA and RRA may or may not be more tolerant of such amino acid replacements, but again, the neurochemical differences must be kept in mind. The MS methodology for analysis discussed in Chapter 7 is the only analytical measurement procedure which is capable of exquisite sensitivity to such structural differences.

3.6 METHOD OF SACRIFICE

The method chosen to sacrifice the animal also plays a significant role, especially as it pertains to the overall metabolism of the cascade of various peptides which are produced under specific metabolic demands and by individual proteolytic enzymes. The microwave process for sacrifice is certainly rapid, especially if appropriately focused with sufficiently high energy. The restriction to this method of sacrifice is the required small size of the animal, limited generally to the rodent. Furthermore, intense temperature in the internal brain structure may liberate lysosomal enzymes. Nevertheless, for a small experimental animal, this methodology is extremely rapid.

Decapitation, especially of the rodent family, is perhaps the most satisfactory method for acquiring brain tissue in a rapid manner. While some cell damage may occur in the plane of decapitation, the other internal brain structures are relatively well-preserved and enzymatic degradation of the peptides of interest is minimal.

Exsanguination is the preferred method of sacrifice, especially when studying the effect of blood levels and blood enzymes. Furthermore, for larger animals such as the dog, this is a very efficient manner of obtaining tissue.

The "emotional state" of the animal at the moment of sacrifice must also be considered (8). While this difficult-to-rationalize or measure factor may not appear to be a significant parameter when working with the lower order animals, it may be a very significant factor in an animal such as the canine model. The emotional state of the animal at the moment of sacrifice may play a significant role in the biochemical and enzymatic disposition of metabolic precursors, especially the larger opioid peptide precursors and the variety of proteolytic enzymes which are brought into play during stress to metabolize the compounds. This is an important fact to remember because generally the objective of the analytical measurement is to provide a "snapshot" of the metabolic profile of that target class of compounds of interest as it existed in an unperturbed intact animal.

3.7 INTERNAL BRAIN STRUCTURES

For neurochemical studies of brain peptides of interest, it is important that a variety of internal brain structures be neuroanatomically identified, neurosurgically removed, and quickly frozen. This identification, separation, and cooling process is done with dispatch to minimize both the time required for tissue procurement and to minimize any potential metabolic conversions. Depending on the particular physiological or neurological parameter(s) to be investigated, any number of the internal brain structures may be excised. For example, if movement disorders (Parkinsonism, Huntington chorea) are of interest, then the basal ganglia are used. For elucidation of nociceptive processes, the spinal cord and thalamus are acquired. For neuroendocrinological interrelationships, the hypothalamus and pituitary are acquired. Figure 3.2 is a drawing to illustrate the dimensional relationships of the various internal brain structures which can be neuroanatomically identified, neurosurgically removed, and stored (9).

3.8 NEUROPEPTIDASE ACTIVITY

Much work has been done recently on studying the enzymes which act upon neuropeptides and their metabolic precursors. The precursors generally contain pairs of the two basic amino acids lysine and arginine which are sensitive to trypsin and/or trypsin-like enzymatic activity. Once this particular enzyme activity hydrolyzes a specific peptide bond in the precursor, carboxypeptidase B activity removes the remaining basic C-terminal amino acid residue. Bioactive peptides such as beta-endorphin, dynorphin, the pentapeptide enkephalins, and substance P are produced by these enzymatic processes and these peptides interact with their respective receptors to

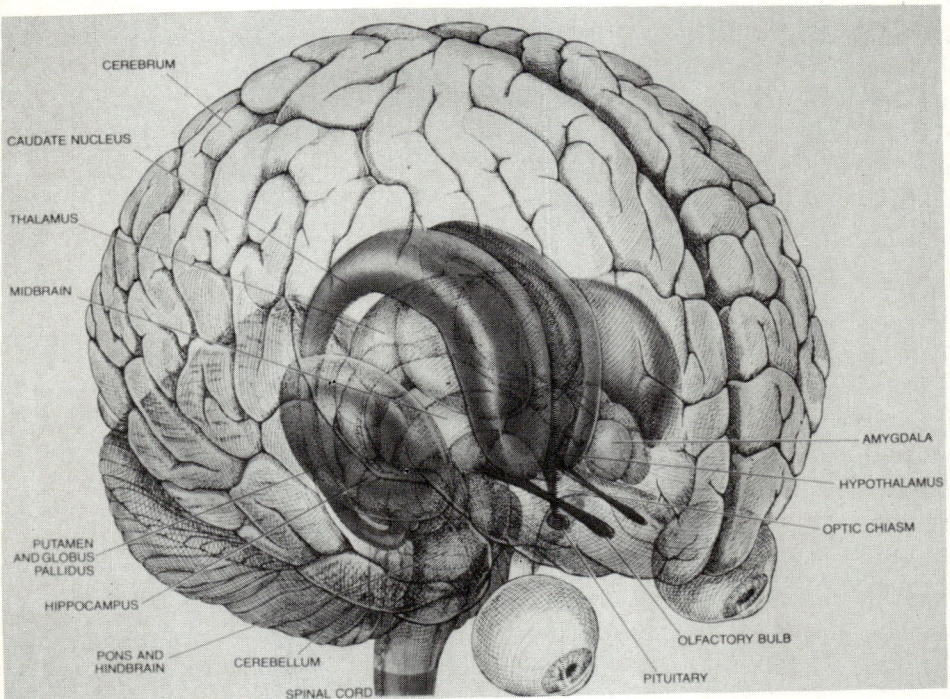

Fig. 3.2. Drawing of internal brain structures (9).

evoke a biologic response (see Chapter 2). Following this receptor interaction, a variety of peptidases acts upon the neuropeptide to rapidly and efficiently remove the active peptide from circulation. These peptidases include aminopeptidases, carboxypeptidases, and specific enkephalinases. Several inhibitors of peptidases may be employed for particular clinical or pharmacological studies. For example, thiorphan is known to be a specific inhibitor of enkephalinase (10), N-ethyl-maleimide, a sulfhydryl reagent, is an irreversible chemical method to inhibit particular enzymes, and both bestatin and bacitracin are protease inhibitors. Alteration of the temperature and pH conditions may also suffice for enzyme inactivation.

Enzymatic activity is an important point to consider when providing biologic tissue for particular studies. For example, if the objective of the experiment is to measure the amount of endogenous peptide by HPLC, MS, BA, or RIA, then it is important that the peptidase activity be minimized to provide an accurate measurement of the peptide in vivo at the moment of sacrifice. On the other hand, if RRA is the objective, it is important to provide a population of biologically active receptors to interact with their appropriate endogenous ligands or with those ligands extracted and purified by HPLC. In the case of RRA, the enzymatic inhibitors may be utilized to retain the peptides to be measured and, for example, irreversible protease inactivators such as N-ethyl-maleimide may not be utilized because these inhibitors may also irreversibly bind to the receptor. Lastly, if the study of precursors is the objective of the study, tissue homogenization must be performed in a neutral aqueous medium and not an acidic organic medium, because use of the latter will precipitate larger peptides.

3.9 SUMMARY

This discussion underscores the importance of being aware of and controlling several crucial experimental parameters which play a role in the acquisition of biological tissue which will be used to provide a peptide-containing fraction and then those endogenous peptides which will be measured in that tissue extract.

REFERENCES

1   M.Z. Zimmermann, Pain, 16 (1983) 109-110.
2   H.T. Jonsson, B.S. Middleditch and D.M. Desiderio, Science, 187 (1975) 1093-1094.
3   H.T. Jonsson, B.S. Middleditch, M.A. Schexnayder and D.M. Desiderio, J. Lip. Res., 17 (1976) 1-6.
4   D.L. Perry and D.M. Desiderio, Prostaglandins, 14 (1977) 745-750.

5   H. Akil, E. Young, S.J. Watson and D.H. Coy, Peptides, 2 (1981) 289-292.
6   N. Brot, J. Werth, D. Koster and H. Weissbach, Anal. Biochem., 122 (1982) 291-294.
7   M.O. Dayhoff (Editor), Atlas of Protein Sequence and Structure -1972, Vol. 5, National Biomedical Research Foundation, Washington, D.C. 1972, 124 pp (plus 418 pp data section).
8   D.B. Carr, Lancet, Feb. 14 (1981) 390.
9   W.J.H. Nauta and M. Feirtag, in The Brain, Freeman, San Francisco, (1979) p. 47.
10   J.-C. Schwartz, B. Malfroy and S. de la Baume, Life Sci., 29 (1981) 1715-1740.

Chapter 4
REVERSED PHASE CHROMATOGRAPHY OF PEPTIDES

4.1 INTRODUCTION

The field of HPLC, especially in the reversed phase (RP) mode, is one of
the most rapidly developing and widely used analytical methods for peptide
purification. While other chromatographic techniques (gel permeation,
electrophoresis, ion exchange) are certainly useful in peptide separation, it
is clear that RP-HPLC is a major contributing factor to recent significant
advances made in several phases of peptide research, including a few selected
areas such as neuroendocrinology, synthesis, and biology. The various aspects
of HPLC involving instrumentation, applications to compound classes, detectors,
and basic theory are discussed in greater depth in other reviews, papers, and
books (1-17). The objective of this chapter is to extract from that published
literature those selected areas of interest which are pertinent to peptide
separation and which are useful to the fields of neuropeptide analysis and MS
quantification. For either a broader or a more detailed analysis of all of
the other pertinent aspects of the field of RP-HPLC, the reader is referred
to the many excellent reference volumes which are currently available (1-17).

4.2 DIFFERENCE BETWEEN REVERSED PHASE AND NORMAL PHASE
CHROMATOGRAPHY

Until a few years ago, normal phase chromatography was the analytical
method which was typically chosen as the basis for chromatographic separation
of peptides. In normal phase chromatography, silica microparticles are used
with a relatively nonpolar mobile phase, and the basis of separation involves
polar interactions between the silica and the peptide of interest. A polar
organic modifier such as methanol is used to elute peptides from the silica
matrix. In RP chromatography, a hydrocarbonaceous group is chemically
bonded to the siliconaceous support material, and the resulting hydrophobic
molecular surface is presented to the compounds to be separated. Those
compounds which possess sufficient hydrophobicity will interact with the
hydrocarbonaceous matrix of the column material and mutual hydrophobic
interactions will occur which are strong enough to preferentially retain
those compounds during elution. While water-soluble, relatively polar
compounds elute with an aqueous wash, those more hydrophobic compounds will

remain attracted within the three-dimensional hydrocarbonaceous matrix of the chromatographic column. Organic modifiers such as acetonitrile have a greater hydrophobic preference for the hydrocarbonaceous surface than the analyte molecules and are used to disrupt the hydrophobic interactions responsible for retaining the peptide to the hydrocarbon surface and to elute peptides of interest. While it might appear surprising that a relatively polar compound such as a peptide possesses sufficient hydrophobicity to participate effectively in RP intermolecular chromatographic interactions, analytical data accumulated over the past several years indicate that RP-HPLC is exquisitely suited for studies requiring high resolution, high sensitivity, and rapid purification of peptides.

The above, brief analysis of the differences between normal and RP chromatography is necessarily concise for the purposes of this book and can not denote all of the molecular, ionic, physical, and dimensional interactions that may participate in chromatographic separation of peptides. For example, the siliconaceous structural backbone of a chromatography column also plays a significant role in RP chromatographic separation processes (18).

4.3 ANATOMY OF A REVERSED PHASE CHROMATOGRAPHY COLUMN

The siliconaceous material utilized for preparation of commercial RP chromatography columns is a silica gel with a pore size ranging from 30 to several hundred microns. Pore size affects the depth of penetration of the peptide and the tortuosity of the path that the peptide takes into and out of the silica gel inner matrix. Pore size also provides an effective physical screen against larger molecular weight proteins.

Details of the commerical processes utilized to prepare RP-HPLC columns are proprietary information. However, generally an alkyl [octyl(C_8) or octadecyl (C_{18})] hydrocarbon chain is chemically bonded to the silica gel matrix via a siloxylic bond which is hydrolytically unstable at pH values more basic than 8. The common abbreviation for octadecylsiloxyl is ODS. Figure 4.1 contains a schematic representation of the silica structure which involves a combination of silicon and oxygen atoms, hydroxyl groups, and siloxane groups.

Our understanding of the behavior of compounds on the columns which are used in LC has increased to the point where several operational features can be rationalized to elucidate the various advantages and disadvantages of using HPLC. These features include the inherent physical and chemical properties of the columns; the limitations to the effective use of those columns; how to optimize their use; and just as importantly, how to avoid their misuse (19).

Hydroxyl Group

Siloxane Group

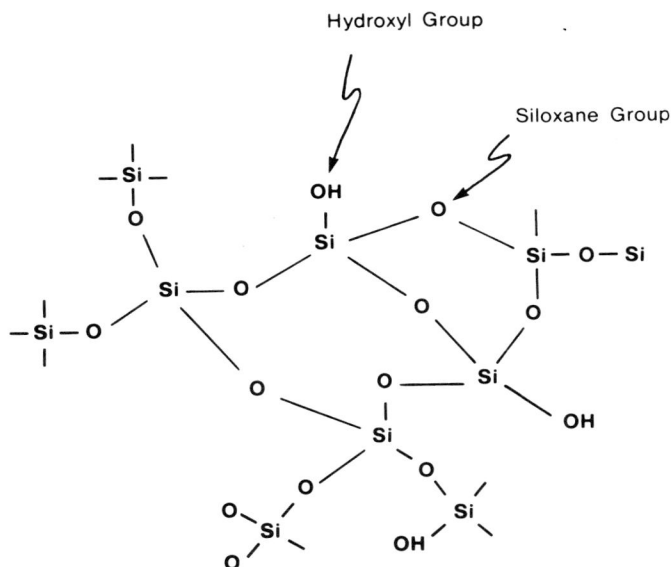

Fig.4.1. Schematic representation of silica structure (adapted from 20).

Historically, three types of LC columns have been used: large porous particles having deep pores with a non-symmetrical (50 micron average) diameter; a porous layer with a pellicular thin film coating of one to two microns depth with shallow pores coated onto a solid core which has approximately 40 microns diameter; and the more recently developed microbead with shallow pores and bead diameter of approximately five microns. The microparticulate beads possess several significant experimental features: a higher value of the critical chromatographic parameter of height of equivalent theoretical plate (HETP); a higher pressure drop; larger sample capacity; and generally, twice the cost. Commercial methods have greatly improved techniques for production of pre-packed microparticulate columns for HPLC, and those columns have more reproducible operating characteristics than comparable columns which were fabricated several years ago. If properly used, a typical microparticulate HPLC analytical column can be used for months for analysis of biological samples without excessive deterioration in chromatographic performance. Longer periods of use are obtained if highly recommended guard columns are also used prior to the analytical column, and if the guard column packing is replaced at appropriate intervals whenever sensitivity and/or resolution decreases.

The most commonly used prepackaged analytical HPLC steel columns have an internal diameter of 4-4.6 mm and a length of 25-30 cm, corresponding to an internal volume of approximately 16 cc. The capacity for a sample is about two to five mg under isocratic conditions for an analytical HPLC column. Microbore columns of one mm internal diameter are now available. Recently, radial compression chromatography columns packaged in fexible polyethylene tubes have also become commercially available. The mechanical holder for these radial compression columns uses a hydraulic fluid such as glycerol to compress the outside of the column and equally exert pressure onto the outside of the pliable polyethylene walls in an effort to eliminate wall-effects during chromatography.

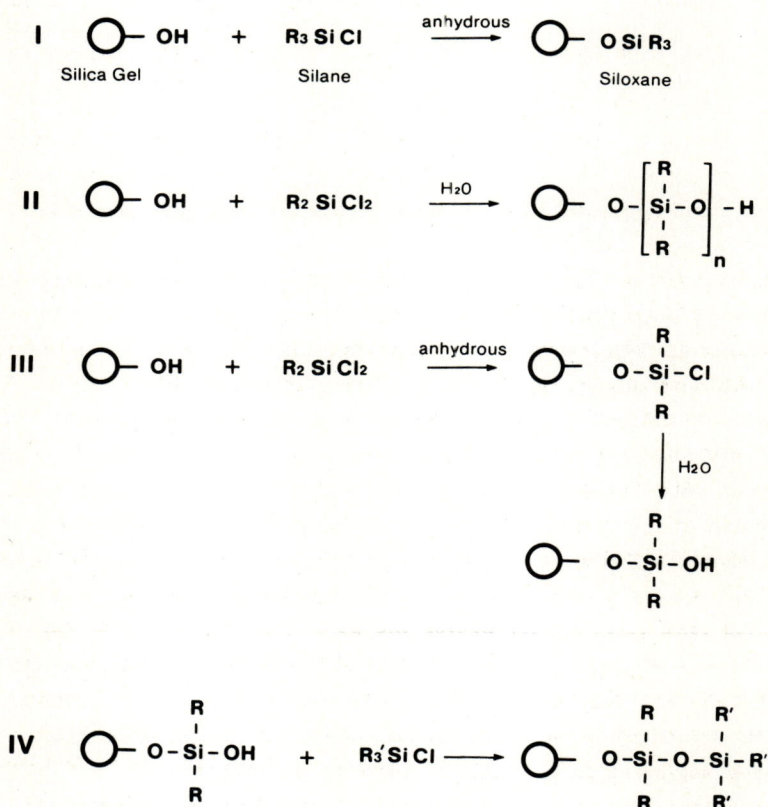

Fig.4.2. Chemical reactions between silica gel and silanes to produce siloxanes (20).

Because of the construction of a microparticulate HPLC column packed with five to ten micron particles, these columns are also excellent filters that will remove any impurities and particulate matter which may be introduced onto the column from a variety of sources including the sample, mobile phase, particles extracted from pump seals and injector core, and any other moving parts of the chromatographic unit. Because of the expense of these analytical columns (hundreds of dollars) and the constant danger of microparticulate matter blocking them, it is highly recommended that all mobile phases be filtered before use through a selected membrane filter (0.45-0.5 micron pores) rather than through a paper filter. Cellulose filters are appropriate for aqueous solutions and fluorocarbon filters for organic solutions. Precolumns are highly recommended to remove particulate matter for a biologic sample and to provide a precolumn volume to effectively saturate the aqueous mobile phase with dissolved silica.

4.4 CHEMICAL SYNTHESIS OF BONDED PHASES

It is important to be aware of the general types of chemical reactions utilized to synthesize commercial bonded reversed phases which are used for packing HPLC columns. Most commercially available bonded phases use preparations of siloxanes, which have the -Si-O-Si-C- bond. Siloxanes are chemically synthesized by reacting the silanol groups of the microparticulate silica gel with either organochloro- or organoalkoxysilanes carried out with mono-, di-, or trichloroorganosilanes (Figure 4.2). Therefore, in general, the recommended operational pH range for these siloxane-bonded silica gels is between pH 2 and 8. If it is found that a particular chromatographic separation must be performed at a very basic pH value, then other types of bonded phase (for example, polystyrene-divinylbenzene) could be used.

The exact, detailed molecular mechanisms involved in RP-HPLC is still a matter of scientific debate. In an effort to rationalize experimental results obtained with RP-HPLC, explanations using "relative phrases" are employed. Polar solutes tend to relatively prefer the polar mobile phase and will therefore elute from an RP-HPLC column before relatively non-polar components which are preferentially retained on the bonded column. The non-polar components in a polar medium of very high cohesive energy density (three-dimensional hydrogen-bonding network) are forced into the hydrocarbon stationary phase. On the other hand, a polar function in a solute molecule will tend to oppose that repulsion from the polar mobile phase. The degree of a solute's retention on the hydrocarbonaceous non-polar column is based primarily on the hydrophobic moiety of that solute. In a similar fashion, if the bonded phase is relatively more hydrophobic, a non-polar molecule will be more greatly attracted.

4.5   MODES OF INTERACTION BETWEEN COLUMN AND SOLUTE

The chromatographic power of separation of HPLC derives from the variety of different modes which are available for analysis. In LC as compared to GC, for example, multiple molecular interactions can occur amongst the mobile phase, the solute, and the stationary phase. There are four modes of analysis available for HPLC:

4.5.1   Liquid-solid chromatography

Liquid-solid (absorption chromatography) generally uses either liquid gel or alumina as the stationary phase absorbant which contains active sites (hydroxyl groups) which will interact with the polar portions of the molecules.

4.5.2   Bonded phase chromatography

Bonded phase chromatography utilizes chemically-bonded phases which are chemically reacted onto a silica gel base. Either partition chromatography or absorption chromatography may be undertaken with these phases. Use of a bonded phase constitutes the most popular form of chromatography utilized today, and it has been estimated that up to 80% of all chromatography done is of the bonded phase type.

4.5.3   Ion exchange chromatography

Ion exchange chromatography utilizes resins or bonded silicas which have ionic groups on their surfaces to attract opposite charges. In addition, a hydrophobic surface may provide a degree of RP separation.

4.5.4   Exclusion chromatography

Exclusion chromatography separates on the basis of molecular size, where larger molecules excluded from a particular pore size will elute first, while smaller molecules which can diffuse into the pores of various sizes will elute later.

4.6 SILICA GEL

Silica gels commonly-used for chromatographic purposes are amorphous, porous solids. These gels can be prepared with a wide range of surface areas and pore diameters where, for bonded phase chromatography, surface areas are usually found with a range of 200-800 $m^2 g^{-1}$ and an average pore diameter range of 50-250 angstroms. It must be remembered that pore diameter can be effectively decreased by the presence of a thick, chemically-bonded phase, where the latter factor affects mass transfer. Chemically, an amorphous silica gel consists of both polysiloxane groups and several types of silanols, where the silanol groups are considered to be those reactive sites which are chromatographically important for adsorption chromatography. These silanol groups are those chemically reactive groups which are covalently bonded to the hydrocarbon chains. A fully hydroxylated surface of silica gel has approximately eight silanol groups nm^{-2}, whereas most commercial silica gels dried at temperatures of less than 200°C have about five silanol groups 100 angstroms^{-2} of surface (20). For a silica gel preparation that has a surface area of 450 m^2 g^{-1}, this level corresponds to three mM silanol groups g^{-1} (seven micromol m^{-2}).

It is important to remember that the pK$_a$ value of a silanol group is 9.5 and it is weakly acidic. Silica does exhibit very slight solubility in a continuously flowing aqueous solution, and this dissolution phenomenon is a fact to remember, especially when MS is utilized as the detector. Sources of potential and observed interferences with MS ionization processes must be eliminated or at least minimized. This solubility phenomenon is particularly troublesome with FD emitters and probably contributes to the emitter breakage problem discussed later (See Chapter 6). Dissolution increases very significantly above pH 8-9 and, because of mass action in a flowing system, solubility of the base silica may also increase. A precolumn guard volume is advantageous in this situation to presaturate the solvent system before the solution enters the analytical column.

Chemical analysis of chemically bonded chromatography phases indicates that coverage ranges from 10-60%. While one would think that the end-capping coverage of all silanol groups is a desirable end point, it must be understood that residual free silanol groups are required to provide "wetting" of the chromatographic phase by the polar modifier of the eluent which will provide more suitable conditions for mass transfer.

Several experimental variables which can be conveniently studied and altered in RP ion-pair chromatography include several aspects of the counter-ion (type, concentration, valence, size, and concentration), pH, type of organic modifier, concentration of organic modifier, rate of change of gradient, column temperature, stationary phase, and number of analytical columns in series.

The relative phase polarity has been more quantitatively explained by utilizing the concept of the Hildebrand solubility parameter, delta. This solubility parameter is a composite factor which incorporates a variety of potential intermolecular interactions that can occur among several experimental components which include the solute, the mobile phase, and the stationary phase. These interactions are based on forces such as dispersion, ole orientation, and hydrogen bonding (1-3, 19,20).

A variety of large pore diameter silicas was coated with N-alkylchlorosilanes and tested for efficacy in protein separation (18). The silica was characterized by studying resolution, load capacity, and desorption. The mechanism of the interaction between the protein and stationary phase was discussed and it was found that the theoretical plate values for small, unretained molecules did not correlate with protein resolution.

4.7 ORGANIC MODIFIERS

The hydrophobic interactions between the peptide of interest and the hydrocarbonaceous packing material must be overcome before a peptide will elute from a column. A variety of organic modifiers has been used in RP-HPLC and, according to increasing elution strength, includes among others: ethylene glycol, methanol, dimethylsulfoxide, ethanol, acetonitrile, dimethylformamide, dioxane, isopropanol, and tetrahydrofuran. A solute will elute earlier if an organic solvent in the above list is replaced by a solvent listed later.

Ultraviolet transparency of the organic modifier is another important physical property to be considered for RP-HPLC of peptides. Optimal peptide bond end absorption occurs in the 190-220 nm wavelength range. While it is true that many biological researchers utilize 254 or 280 nm for peptide and protein monitoring, greater detection sensitivity is obtained at the lower UV wavelengths, because lower wavelengths correspond to peptide bond absorption and not to a limited number of particular amino acid residues.

Whenever either the aqueous buffer or organic modifier has a relatively high background level absorption at the UV wavelength which is being monitored, an electronic circuit can compensate for that high background ultraviolet absorbance level, or a set of dual detector cells can be used with only buffer:organic modifier stored in or flowing through the unused cell.

4.8 BUFFERS

A number of buffers has been developed over the past few years and several are very effective for peptide separation utilizing RP-HPLC. A buffer is used either to suppress or to control the dissociation of ionizable groups which are present in the amino acid residues and terminal groups in a peptide. The pK$_a$ values of individual amino acid residues, the free N-terminus, and free C-terminus must all be considered in any detailed mathematical analysis of ionization suppression. Furthermore, the chemical susceptibility of the ODS bond to hydrolysis must also be considered, because the pH of the buffer must remain in the range of 2-8 to avoid hydrolysis of the packing material.

Triethylamine:phosphoric acid (TEAP) is an excellent buffer developed for peptides (21) and, while the TEAP buffer possesses appropriate UV transmission properties and displays high resolution and high sensitivity capabilities for peptide analysis, especially during gradient analysis, this non-volatile buffer limits the direct use of certain subsequent detection methods such as MS, RRA, BA, and RIA. An additional salt removal step is required whenever TEAP must be used as an HPLC buffer.

TEAP

A volatile triethylamine:formic acid (TEAF) buffer was developed from formic acid (40 mM) which is titrated (pH 3.14) with distilled triethylamine (22). The TEAF buffer possesses a higher UV background than TEAP, and this background level may be compensated for by the electronic circuitry of commercial HPLC UV detectors.

Volatile perfluoroalkanoic acid buffers have been used with advantage for peptide separations (23). The increased hydrophobicity of the perfluoroalkyl groups plays a role in peptide retention in the RP mode.

4.9 LOW RESOLUTION MINI-CHROMATOGRAPHIC COLUMNS

Small chromatography columns containing ODS RP packing material (ca. 0.5 g) are commercially available and have assumed a level of popular usage. Use of these small columns is advocated (24) for preliminary low resolution purification of tissue extracts to preferentially retain peptides while eluting salts and other water-soluble polar compounds. The peptide-rich fraction retained on the ODS surface is conveniently eluted with acetonitrile (80%) and subsequently separated in the high resolution mode with an analytic HPLC column.

The C_{18} hydrophobic sidechains in a mini-column are generally prepared before use for low resolution chromatographic separations to efficiently maximize the capacity and resolution of the column. The mini-column is washed first with organic solvents to extend the C_{18} side-chains like a "bristle" to maximize the surface area which is available for hydrophobic interactions of the ODS packing with the hydrophobic regions of the ion-pair:peptide complex (25).

Mini-columns are also used as a preliminary purification step before RIA measurement of enkephalins in a series of rat brain tissue extracts (26); before HPLC analysis of insulin metabolites (27); and for prepuri- fication of neuropeptide extracts of CSF and brain tissue extracts as a

preliminary to MS (28), RIA (29), and RRA (30) measurement of endogenous
ME and LE.

A relatively simple, fast, inexpensive, and reproducible method for
separation of enkephalins and endorphins on RP Sep-Pak cartridges is
described (31). The Sep-Pak cartridges contain about 800 microliters of C_{18} RP
packing in a void volume of about 400 microliters. Cartridges are prepared
first by washing with methanol (2 ml), de-ionized water (5 ml), and then
formic acid:pyridine buffer (5 ml, 0.5 M, pH 3-4). A step-gradient of
n-propanol in a formate:pyridine buffer is used to elute sample (one-two
drops sec^{-1}, one ml fractions). A mixture of ^{125}I (0% n-propanol),
alpha-endorphin (420 micrograms at 10% propanol), gamma-endorphin (340
micrograms at 20% propanol), and beta-endorphin (19 microcuries at 30%
propanol) was readily resolved. ME and LE were readily separated at 8% and
15% propanol, respectively. It was noted that ME sulfoxide elutes at 5%
propanol. In a metabolic study, tyrosine (0% propanol), des-tyr-ME (5%
propanol), and ME (10% propanol) are readily separated. Reproducibility was
noted were elution of tritiated ME from one run was superimposed on ME from
another sample which was run on a different cartridge.

4.10 PEPTIDE SEPARATIONS

Brain peptides have different relative hydrophobicities and
consequently, different retention times are observed on a RP-HPLC column
whenever perfluorinated alkanoic acids (for example, trifluoroacetic and
heptafluorobutyric acids) are utilized as the buffer. These differential
hydrophobic effects can be used to advantage in both the isolation and
purification of endogenous peptides from biologic sources (32). Two major
forms of adrenocorticotropic hormone (ACTH) from the rat anterior pituitary
were isolated using only RP chromatographic methodology, where a method for
consistently and completely inhibiting peptidase activity was employed by the
use of an extraction medium with an acidic pH (1 M HAc at 4°C). For the first
time, these ACTH peptide molecules were isolated from rat pituitary extracts
and identified. The HPLC-grade water used in this study was prepared from
de-ionized, glass-distilled water which was subsequently passed at a slow rate
of elution under gravity over a small bed of ODS silica which was taken from one
C_{18} Sep-Pak. Pituitaries were removed following decapitation, dissected into
anterior and neurointermedullary lobes, and rapidly (within one minute)
homogenized (4°). Supernatants were combined and passed five times through
Water C_{18} Sep-Paks. Plasma was obtained and three-ml portions extracted by
passing ten times through cartridges. These "multi-passages" through Sep-Paks

were used to provide sufficient capacity of the ODS RP packing. Recoveries of the various forms of peptide were very high. It was interesting to note that seven synthetic peptides were separated by HPLC acetonitrile gradients in trifluoroacetic (0.01 M) versus heptafluorobutyric acid (0.01 M). While several differences were observed between the two HPLC chromatograms, the most important difference was that the order of elution of the peptide standards differs between the two chromatograms. Because the acetonitrile gradient is identical in both cases, the difference in chromatographic elution behavior is clearly due to only the hydrophobicity of the counter-ion, the perfluorinated carboxylic acid. The retention times of all peptides are greater in the heptafluorobutyric acid chromatogram compared to the smaller carboxylic acid. Peptides with a greater number of basic groups at pH 2 show large increases in retention times when comparing the heptafluorobutyric to the trifluoroacetic acid system, while in contrast, peptides which contain relatively few basic groups show smaller increases.

A variety of workers has recently demonstrated that peptides can be readily separated by RP-HPLC, and the most successful chromatographic separation methods depend on the use of an ionic-modifier or ion-pairing reagent in the solvent system to provide good separation. These types of modifiers can be conveniently divided into two large groups; acids at pH 2-3 (such as trifluoroacetic, phosphoric, and hydrochloric acids) whereas the second broad class includes buffered salts (such as triethylammonium phosphate, ammonium acetate, or triethylamine formic acid).

[handwritten margin notes: ion-pairing reagent; acids or salts]

Two major forms of the immunoreactive neuropeptide ACTH found in the rat neurointermediary pituitary were underivatized corticotropin-like intermediate peptide (CLIP) and O-phospho-^{31}seryl-CLIP, with the phosphorylated form predominating by a 2:1 ratio (33). The main forms of immunoreactive alpha-MSH found were N-acetyl-^{1}ser- alpha-MSH, and N, O-diacetyl-^{1}ser- alpha-MSH, with the diacetyl form predominating by a 9:1 ratio. It is significant to note that all four of these neuropeptides were purified by only two RP-HPLC steps. Phosphorylation of the ^{31}ser-CLIP molecule is a post-translational modification which most probably serves as a mechanism to mask the recognition site for glycosylation of the ^{29}asn residue.

RP-HPLC techniques were utilized in a study of the in vitro incorporation of ^{32}P into ACTH-related peptides by rat pituitary tissue (32). HPLC was used to study alpha-N-acetyl-beta-endorphin$_{1-26}$ from the rat pituitary neurointermediate lobe. Phosphorylation of pro-adrenocorticotropin and endorphin-derived peptides was studied (34).

4.11 HIGH RESOLUTION ANALYTIC REVERSED PHASE HIGH PERFORMANCE LIQUID CHROMATOGRAPHY COLUMNS

The last step recommended before any analytical measurement (MS, RIA, RRA, BA) of peptides is HPLC in the RP mode. RP-HPLC provides high sensitivity, high resolution, and high speed of peptide separation. Synthetic solutions of the tetradecapeptide somatostatin are quantified down to the 500 fmol (0.9 ng) level with good straight-line statistics (Fig. 4.3) in a study to demonstrate high sensitivity and resolution (22). The RP-HPLC resolution of peptide mixtures is high (Fig. 4.4). Many longer peptides are differentiated from their homologs, where one residue is replaced with another which differs only by one methylene unit. For example, two pentapeptides differing by only one methyl group are readily separated (35). LE is separated by seven minutes from its homolog (TyrAlaGlyPheLeu.)

4.12 HIGH PERFORMANCE LIQUID CHROMATOGRAPHY DETECTORS
4.12.1 Ultraviolet detectors

The detector most utilized for conventional column chromatography in general and HPLC in particular is the ultraviolet detector. Wavelengths utilized for a typical UV detection system are 280 nm for proteins and 254 nm for nucleic acids. The protein-sensitive wavelength (280 nm) monitors mainly the W amino acid residue and, with decreasing levels of sensitivity, Y and then F. While 280 nm is an appropriate wavelength for monitoring larger proteins because of the probability that a W, Y, or F residue is present, the sensitivity of monitoring shorter peptides optimizes at 190-210 nm due mainly, and in some cases only, to peptide bond absorbance and not to the presence of a W, Y, or F residue. This difference in sensitivity of different wavelengths is an important point to consider because, with diminishing length of a protein or larger peptide, the probability also decreases that a W, Y, or F amino acid residue will be present in the peptide molecule. It is for this statistical reason that the peptide amide bond is most appropriately monitored at 200 nm for peptide elution instead of only that wavelength corresponding to a selected amino acid side-chain. For peptide monitoring at 200 nm, a high molar absorbtivity (10^5 for ME) is achieved leading to femtomole (fmol, 10^{-15} mole) sensitivity of peptide detection (22). On the other hand, monitoring at 280 nm for an amino acid could have orders of magnitude less molar absorbtivity.

While most HPLC detectors have a variable wavelength ultraviolet detector which is capable of operating at approximately 200 nm, it has been found that there is an increase in absorbance at 187 nm which represents a three-fold

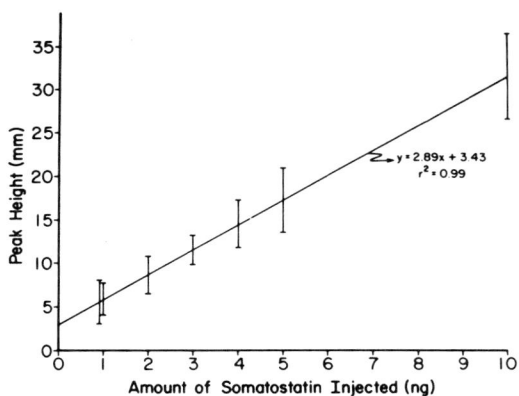

Fig.4.3. Straight-line plot of peak height vs. amount of synthetic
somatostatin which is injected onto a RP-HPLC column (22).

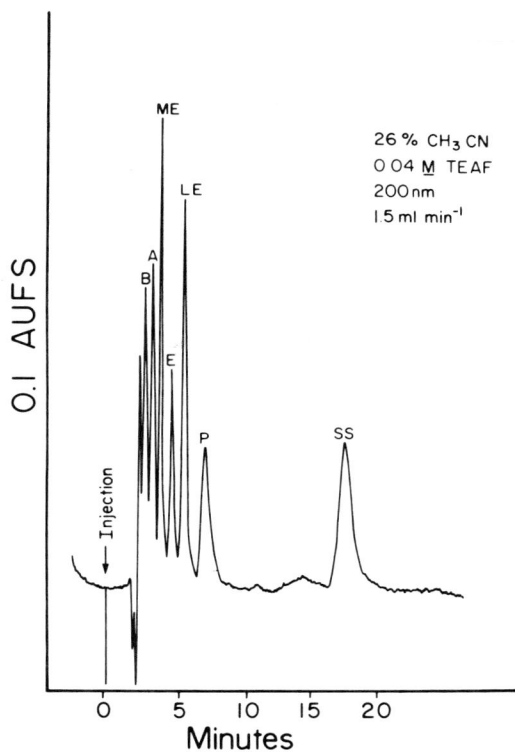

Fig.4.4. Isocratic RP-HPLC chromatogram of a mixture of synthetic peptides.
B = bradykinin; A= angiotension II; ME = methionine enkephalin;
E = eledoisin-related peptide; LE = leucine enkephalin; P =
substance P; and SS = somatostatin (22).

increase versus 210 nm and four-fold versus 205 nm (36). It was found in another study that anions (hydroxide, chloride) and oxygen interfere because they also absorb at 190 nm (37). Either Sephadex G25 or dialysis against fluoride ion was utilized to remove chloride ions.

A recently developed commercial photodetector allows simultaneous data acquisition of intensity data from light transmission at wavelengths between 190-600 nm (38). The photodiode array used in this detector contains 211 photosensitive diodes, each of which has its own individual storage capacitor connected in parallel to a microprocessor, where photocurrent resulting from light striking the diode causes the capacitor to discharge towards zero potential. The amount of discharge relates to the amount of light that fell on the diode. Serial readout of the array is accomplished by a combination of digital techniques, analog-to-digital convertors, and a microprocessor. A full absorbant spectrum from 190-600 nm is acquired in ten msec.

4.12.2 Detectors based on fluorescence

It is possible to take advantage of the intrinsic fluorescence of specific amino acid residues such as tryptophan. An elegant separation scheme is published using differential fluorescence and UV detection modes to determine the presence of specific amino acid residues in a peptide (39). Furthermore, it is possible to chemically derivatize the N-terminus or C-terminus of a peptide with a fluorogenic group and then use a fluorescence detector for quantification. Fluorescamine is an excellent example of this procedure.

In a detector based on fluorescence, a compound is excited at one specific wavelength and a second emission wavelength is monitored. The excitation wavelength is produced and the emission wavelength monitored conveniently by using appropriate filters, which can have either a broad or narrow band-pass wavelength. The advantages that fluorescence detection offers to peptide analysis include both a low noise-level and a high signal-to-noise ratio. Again, however, caution must be exercised regarding statements and conclusions involving molecular specificity when using only a fluorescence detector to monitor peptides eluting from an HPLC column.

4.12.3 Electrochemical detectors

This type of HPLC detector evolved from electroanalytical chemistry. Specific electroactive species in a molecule are amenable to electrochemical reduction and/or oxidation (redox) reactions and are appropriate for electroanalytical measurement. Commercially available electrochemical detectors are available for HPLC detection. The putative specificity of an electrochemical detector derives from the fact that an electroactive species has a specific redox half-wave potential. All other electroactive species which may be reduced or oxidized at other potentials are not electroactive at that selected potential. These electrochemical potentials are conveniently available from the published electroanalytical literature. The electrical current (generally, in nanoamperes) produced at the redox potential of the electroactive species is monitored and can be related to the amount of electroactive species present (40). Again, care must be exercised regarding statements on molecular specificity.

Electroactive species present in peptides can be readily quantified by utilizing the combination of RP-HPLC plus an electrochemical detector (40). One intriguing possibility is the use of implanted microelectrodes to follow neurochemical processes within a cell.

4.13 REVERSED PHASE HIGH PERFORMANCE LIQUID CHROMATOGRAPHY OF BIOLOGIC PEPTIDES

While RP-HPLC has been utilized for a variety of compounds, this chapter and the entire book focus on peptide analysis in particular. Several books are available for HPLC analysis of other compounds (1-3, 6-8, 12).

The first detailed analysis of HPLC of peptides included a study of recovery, sensitivity, temperature effects, resolution, and other experimental parameters (21). RP-HPLC was used to study biologically important brain peptides (41). Since that time, many publications have appeared utilizing the technique.

RP-HPLC effectively separates enkephalin peptides (41a). For example, the following series of figures are presented to illustrate the isocratic RP-HPLC of enkephalins which are extracted from a biologic tissue (hypothalamus) in Fig. 4.5 and a fluid (CSF) in Fig. 4.7. Also shown are corresponding chromatograms (Figures 4.6 and 4.8, respectively) to which the stable isotope-incorporated peptide internal standard has been added for MS measurement (Chapter 7). The chromatograms demonstrate that these extracts are relatively clean in the isocratic mode when monitored with UV detection. However, RRA and RIA of these extracts indicates nanogram amounts of LE and ME (Chapter 5). The "cleanliness" of the HPLC chromatograms may be misleading in that only UV chromophores are detected. It is possible that non-absorbing material also elutes from the HPLC column and these compounds may interfere with several types of subsequent assay methodologies.

While the basic molecular processes involved in RP-HPLC relating to the interaction of the ion-pair formed between peptide and buffer with the hydrocarbonaeous column material are not yet completely described, detailed mathematical and experimental data have been analyzed (42). The theory of ion-pair chromatography is analyzed as well as those factors which participate in controlling the capacity factor, k'.

The effect of comparing phosphoric acid (0.1%, pH 2.5) and acetic acid (0.1%, pH 4) on the retention times of various peptides demonstrate that a significant reduction in retention time is obtained (43). A series of different ion-pairing reagents demonstrates changes in selectivity of a RP system for a range of peptides from tripeptides to heptapeptides by utilizing a series of ion-pairing reagents which includes phosphoric acid, sodium hexanesulfonate, and sodium dodecylsulfate (44).

An amphoteric molecule such as a peptide may participate readily in either hydrophilic or hydrophobic ion-pairing processes and poor chromatographic resolution may be observed if sufficient ion-pairing is not performed. These complications are not surprising if one considers the number of complex ionic equilibria in which these amphoteric compounds can participate (45). One problem encountered occasionally is that smaller peptides are sufficiently polar and are retained on a hydrocarbonaceous column. In those cases, a series of perfluoroalkanoic acids may be utilized to extend the retention time of that peptide (46). For certain tripeptide solutes, retention time is increased when the mobile phase is changed from trifluoroacetic to pentafluoropropionic acid, then increased further with heptafluorobutyric acid. This effect of increasing retention time may be rationalized due to the lipophilicity of the perfluoroalkyl groups which result in a decreased polarity of the ion-pair formed between that peptide solute and the perfluoroalkanoic acid present in the mobile phase.

RP-HPLC conditions are optimized for the cyclic tetradecapeptide somatostatin (47). A variety of phosphate, acetate, and sulfonic acids are utilized. Organic modifiers were tested and included acetonitrile and various alcohols. Temperature effects and buffer concentrations were changed to optimize the retention time of somatostatin. While sensitivity of detection permitted determinations down to 10-20 ng of somatostatin, this level of sensitivity must be compared to that sensitivity observed for synthetic somatostatin solutions down to 900 pg, corresponding to 615 fmol, with TEAF buffer (22).

The peptide hormone oxytocin and seven of its diastereoisomers were studied and their chromatographic behavior on RP-HPLC columns rationalized (48). Effects due to solvent, pH, and salt concentration as well as the different solvent systems (10% tetrahydrofuran-ammonium acetate, or 18% acetonitrile-ammonium acetate) were studied. In all cases, the

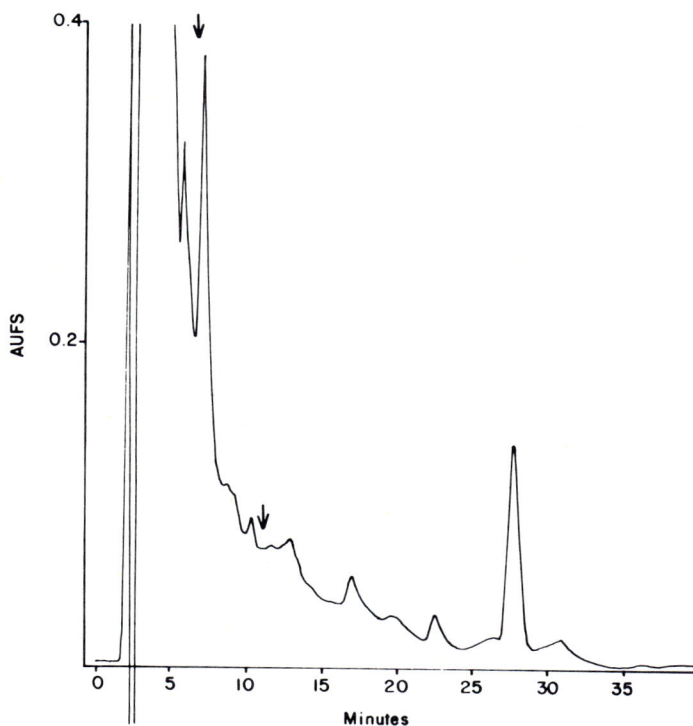

Fig.4.5. RP-HPLC of the peptide-rich fraction extracted from canine hypothalamus. Arrows indicate retention time of ME and LE, respectively (41a).

diastereoisomers eluted later than oxytocin and, in conjunction with nuclear magnetic resonance (NMR) studies, a rationale was proposed which stated that conformation of the oxytocin peptide chain, the hydrophobicity of the peptide, and hydrophilicity of the mixture of aqueous buffer and organic modifier all play roles in the retention of these hormones on the hydrocarboneous surface.

Pairs of diastereoisomers were also studied by HPLC (41).

Two chromatographic techniques, ion-exchange chromatography and RP-HPLC, are complementary techniques used in conjunction with RIA and BA for known peptides (49). HPLC was utilized on acid-boiled crude rat brain extract and, in conjunction with RIA, demonstrated that the enkephalins were present in the eluate, even though a high UV background was noted. While these authors utilize 280 nm for determination of peptides, other workers have shown that the 210 nm range is more sensitive towards peptides (22).

Yields of peptide eluted from the HPLC column are excellent and, for
^{125}I-labeled VIP, demonstrate that recoveries exceed 85%.

A simpler method has been developed for purification of GI neuropeptides
from relatively small amounts of GI tissue (50). The experimental sequence of
sequential absorption on gel filtration, ion-exchange chromatography, and
HPLC without lyophilization produces good yields for somatostatin, bombesin,
and VIP immunoreactivity. The authors note that rigorous maintenance of acidic
conditions is required to achieve optimal recoveries of these basic peptides.
Acidity ensures that the carboxylate, guanidinium, and amino group (Lys and
N-terminus) remain protonated. VIP was first purified from hog intestine,
somatostatin from ovine hypothalamus, and bombesin from frog skin. Speculation
has increased that these peptides may function as neurotransmitters (Table 2.3).

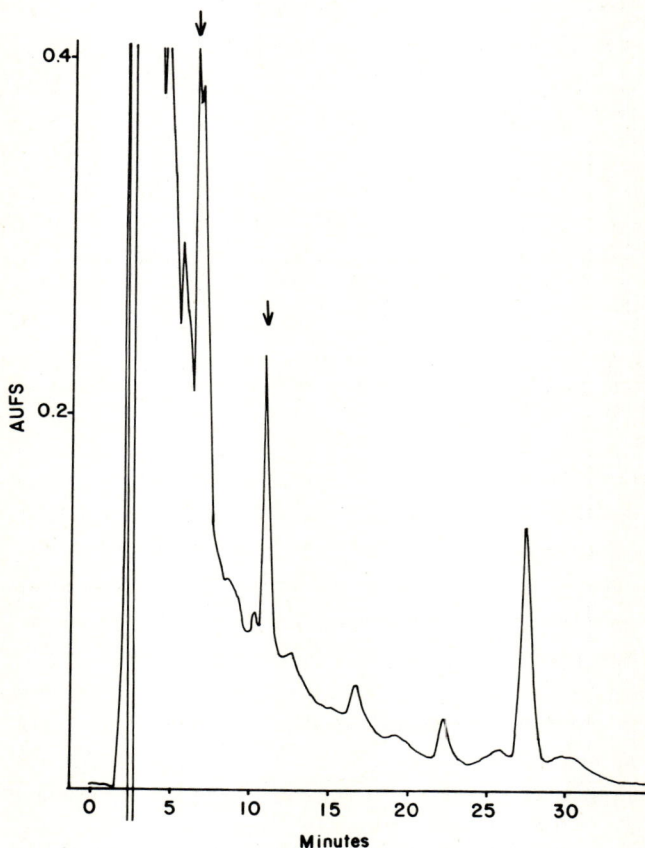

Fig.4.6. RP-HPLC of the peptide-rich fraction extracted from canine
hypothalamus which is spiked with ^{18}O-ME and ^{18}O-LE (denoted
by arrows, respectively) (41a).

The observation that trifluoroacetic acid enhanced the partition of opioid peptides into butanol during aqueous-organic phase extractions led to the use of trifluoroacetic acid as a counter-ion for ion-pair partition RP-LC. Trifluoroacetic acid is an effective organic modifier because it provides decreased nonspecific absorption of opioids and opioid peptides onto column matrices while decreasing elution time without loss of resolution (51).

4.14 OTHER ASPECTS

Recent developments in HPLC involve microbore HPLC where the column diameter is much smaller (one mm), flow rates are reduced, resolution is higher, and less organic solvent is utilized. These attributes make microbore techniques amenable to an on-line combination with an MS detector (Chapter 8).

A reference book (13) discusses the observation that individual amino acids and other compounds possess inherent physiochemical properties which cause that compound to distribute between a polar and less-polar solvent. This

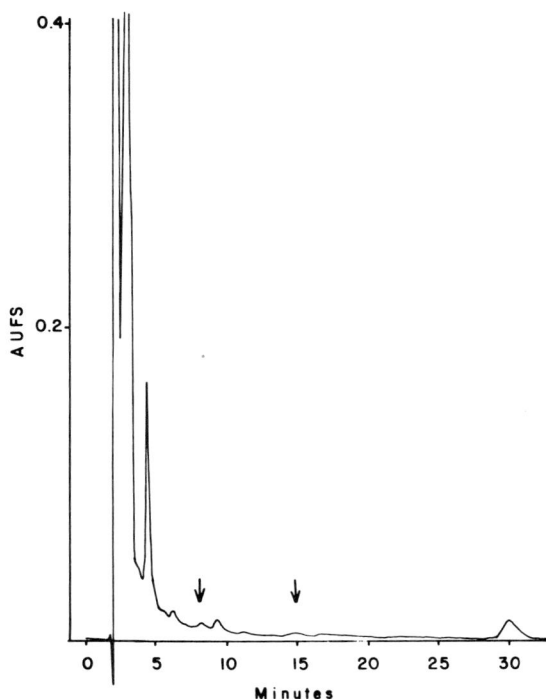

Fig. 4.7. RP-HPLC of the peptide-rich fraction extracted from canine CSF. Arrows indicate retention time of ME and LE, respectively (41a).

distribution pattern is unique to each compound and relates to the intrinsic hydrophobicity of the compound relative to a selected solvent system, and an empirical set of distribution constants has been derived experimentally. For a peptide, an integrated hydrophobicity constant (Rekker) can be calculated by summing the fragmental hydrophobicity constant of each constituent amino acid. A large number [32] of polypeptide hormones and fragments, and nine proteins, ranging in molecular weight from 8,000 to 65,000 daltons, were studied by using alkylsilyl-bonded microparticulate porous silica packings (52).

The retention orders of smaller peptides (less than 15 amino acid residues) in general correlates with the integral of the Rekker fragmental constants for the more strongly hydrophobic amino acid residues. However, larger polypeptides deviated from this ordering and limited the predictive value of the integrated Rekker constants. Another paper suggests that, for peptides containing up to 20 residues, retention is primarily due to a partition

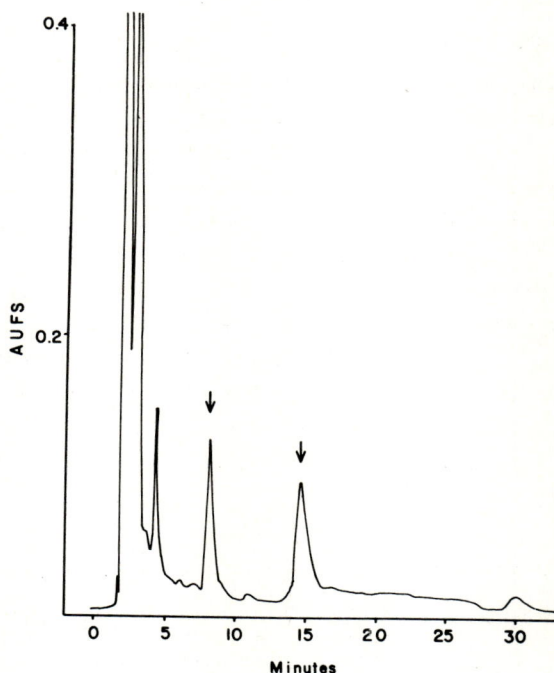

Fig. 4.8.   RP-HPLC of the peptide-rich fraction extracted from canine CSF which is spiked with ^{18}O-ME and ^{18}O-LE (denoted by arrows, respectively) (41a).

process that involves all of the amino acid residues (53). These data suggest
that retention of peptides containing up to 20 residues may possibly be
predicted only on the basis of their amino acid composition. Other data show
that the hydrophobicity, which relates to the RP-HPLC elution time of a peptide,
may relate to the sum of the fragmental hydrophobicity constants (28, 53). This
is an interesting phenomenon with perhaps only limited potential analytical
utility however, because in principle, the RP-HPLC retention time cannot be
used to unambiguously establish the amino acid sequence of longer peptides.
Nonetheless, it is interesting to note that this type of experimental approach
is comparable to a GC-MS method for peptide analysis (54), where peptides are
chemically reduced to their corresponding polyamino alcohols which are converted
to trimethylsilyl derivatives. The GC retention times of the
trimethylsilylated-polyamino alcohols correlate with the sum of the individual
fragmental GC retention times of the individual amino acids for peptides
containing up to six amino acid residues.

4.15 SUMMARY

Much of the increased research activity in biologically active peptides
may be attributed directly to developments in RP-HPLC methodology.
RP-HPLC effectively and efficiently separates peptides with high resolution,
high speed, and high recovery. It is crucial to review the basic concepts of
RP-HPLC and to understand how to take advantage of several experimental
features of RP-HPLC. Ion-pairing, buffers, organic modifiers, and
hydrocarbonaceous surfaces must be studied, understood, and manipulated
experimentally to increase the efficiency of analytical separations of
peptides. The types of detectors are reviewed with a view towards
understanding the claims that can be made regarding the critical parameter
of molecular specificity.

REFERENCES

1   C. Horvath (Editor) High Performance Liquid Chromatography: Advances
     and Perspectives, Vol. 1, Academic Press, New York, 1980, 330 pp.
2   C. Horvath (Editor), High Performance Liquid Chromatography: Advances
     and Perspectives, Vol. 2, Academic Press, New York, 1980, 340 pp.
3   C. Horvath (Editor), High Performance Liquid Chromatography: Advances
     and Perspectives, Vol. 3, Academic Press, N.Y., 1983 230 pp.
4   J.C. Giddings, E. Grushka, J. Cazes, and P.R. Brown (Editors),
     Advances in Chromatography, Vol. 18, Marcel Dekker, New York, 1980,
     292 pp.
5   J.C. Giddings, E. Grushka, J. Cazes, and P.R. Brown (Editors), Adv.
     Chromatogr., Vol. 22, Marcel Dekker, NY, 1983, 323 pp.

6   G. Hawk, P.B. Champlin, H.C. Jordi and D. Wenke (Editors), Biological/
     Biomedical Applications of Liquid Chromatography, Vol. 10, Marcel
     Dekker, New York, 1979, 736 pp.
7   G. Hawk, P.B. Champlin, R.F. Hutton, H.C. Jordi and C. Mol (Editors),
     Biological/Biomedical Applications of Liquid Chromatography II, Vol. 12,
     Marcel Dekker, New York, 1979, 504 pp.
8   G. Hawk, P.B. Champlin, R.F. Hutton and C. Mol (Editors), Biological/
     Biomedical Applications of Liquid Chromatography III, Vol. 18, Marcel
     Dekker, New York, 1981, 420 pp.
9   A. Zlatkis, R. Segura and L.S. Ettre (Editors), Advances in
     Chromatography-1981, Chromatography Symposium, University of
     Houston, Houston, Texas, 591 pp.
10   A. Zlatkis, L.S. Ettre and C.F. Poole (Editors), Advances in
     Chromatography 1982, Chromatography Symposium, University of
     Houston, Houston, Texas, 448 pp.
11   A.M. Lawson, C.K. Lim and W. Richmond (Editors), Current Developments
     in the Clinical Applications of HPLC, GC and MS, Academic Press, New
     York, 1980, 301 pp.
12   P.R. Brown, High Pressure Liquid Chromatography: Biochemical and
     Biomedical Applications, Academic Press, New York, 1973, 202 pp.
13   C. Tanford, The Hydrophobic Effect - Formation of Micelles and
     Biological Membranes, Wiley, New York, 1980, 233 pp.
14   L.R. Snyder and J.J. Kirkland, Introduction to Modern Liquid
     Chromatography, Wiley, New York. 1974, 534 pp.
15   L.R. Snyder and J.J. Kirkland, Introduction to Modern Liquid
     Chromatography, 2nd Ed., Wiley, New York, 1979, 863 pp.
16   E. Heftmann (Editor), Chromatography, A Laboratory Handbook of
     Chromatographic and Electrophoretic Methods, Van Nostrand-Reinhold,
     New York, 1975, 969 pp.
17   W.W. Yau, J.J. Kirkland and D.D. Bly, Modern Size-Exclusion Liquid
     Chromatography, Wiley, New York, 1979, 476 pp.
18   J.D. Pearson, N.T. Lin and F.E. Regnier, Anal. Biochem., 124 (1982)
     271-230.
19   R.E. Majors, in L.J. Marton and P.M. Kabra (Editors), Liquid
     Chromatography in Clinical Analysis, The Humana Press, Clifton, N.J.,
     1981, pp. 51-94.
20   R.E. Majors, in High Performance Liquid Chromatography, Vol. 1, Academic
     Press, 1980, pp. 75-111.
21   J.E. Rivier, J. Liq. Chromatogr.,1 (1978) 343-366.
22   D.M. Desiderio and M.L. Cunningham, J. Liq. Chromatogr., 4 (1981)
     721-733.
23   H.P.J. Bennett, C.A. Browne and S. Solomon, Biochem., 20 (1980)
     4530-4538.
24   H.P.J. Bennett, C.A. Browne, P.L. Brubaker and S. Solomon, in G.L.
     Hawk (Editor), Biological/Biomedical Applications of Liquid
     Chromatography III, Marcel Dekker, Inc., New York, 1981,
     pp. 197-210.
25   J.O. Whitney and M.M. Thaler, J. Liq. Chromatogr., 3 (1980) 545-556.
26   P. Angwin and J.D. Barchas, J. Chromatogr., 231 (1982) 173-177.
27   F.B. Stentz, R.K. Wright and A.E. Kitabchi, Diabetes, 31.(1982)
     1128-1131.
28   D.M. Desiderio, S. Yamada, F.S.Tanzer, J. Horton and J. Trimble, J.
     Chromatogr., 217 (1981) 437-452.
29   F.S. Tanzer, D.M. Desiderio, C. Wakelyn, and J. Walker, J. Dent. Res.,
     submitted.
30   D.M. Desiderio, H. Onishi, H. Takeshita, F.S. Tanzer, C. Wakelyn, J.A.
     Walker, Jr., and G. Fridland, J. Neurochem., submitted.

31   D.D. Gay and R.A. Lahti, Int. J. Pept. Prot. Res., 18 (1981) 107-110.

32   H.P.J. Bennett, C.A. Browne and S.Solomon, Proc. Natl. Acad. Sci. USA, 78 (1981) 4713-4717.

33   C.A. Browne, H.P.J. Bennett and S. Solomon, Biochem., 20 (1981) 4538-4546.

34   B.A. Eipper and R.E. Mains, J. Biol. Chem., 257 (1982) 4907-4915.

35   D.M. Desiderio, S. Yamada, F.S. Tanzer, J. Horton, and J. Trimble, J. Chromatogr., 217 (1981) 437-452.

36   A.H. Woods and P.R. O'Bar, Science, 167 (1970) 179-181.

37   M.M. Mayer and J.A. Miller, Anal. Biochem., 36 (1970) 91-100.

38   J.C. Miller, S.A. George, and B.G. Willis, Science, 218 (1982) 241-246.

39   C.T. Wehr, L. Correia and S.R. Abbott, J. Chromatogr. Sci., 20 (1982) 114-119.

40   P.T. Kissinger, C.S. Bruntlett and R.E. Shoup, Life Sci., 28 (1981) 455-465.

41   J. Rivier, R. Wolbers and R. Burgus, in M. Goodman and J. Meienhofer (Editors), Peptides - Proc. Fifth Amer. Pept. Symp., Wiley, New York, 1977, p. 52-55.

41a   D.M. Desiderio, M. Kai, F.S. Tanzer, J. Trimble, and C. Wakelyn, J. Chromatogr., in press.

42   M.T.W. Hearn, in J.C. Giddings, E. Grushka, J. Cazes, and P.R. Brown (Editors), Advances in Chromatography, Vol. 18, Marcel Dekker, New York, 1980, pp. 59-100.

43   W.S. Hancock, C.A. Bishop, R.L. Prestidge, D.R.K. Harding and M.T.W. Hearn, J. Chromatogr., 153 (1978) 391-398.

44   W.S. Hancock, C.A. Bishop, L.J. Meyer, D.R.K. Harding and M.T.W. Hearn, J. Chromatogr., 161 (1978) 291-298.

45   W.S. Hancock, C.A. Bishop, R.L. Prestidge, D.R.K. Harding and M.T.W. Hearn, Science, 200 (1978) 1168-1170.

46   D.R.K. Harding, C.A. Bishop, M.F. Tarttelin and W.S. Hancock, Int. J. Pept. Res., 18 (1981) 214-220.

47   M. Abrahamsson and K. Groningsson, J. Liq. Chromatogr., 3 (1980) 495-511.

48   B. Larsen, B. L. Fox, M.F. Burke and V.J. Hruby, Int. J. Peptide Res., 13 (1979) 12-21.

49   H.R. Morris, A.T. Etienne, A. Dell and R. Albuquerque, J. Neurochem., 34 (1980) 574-582.

50   J. Reeve, T. Yamada, P. Chew and J.H. Walsh, J. Chromatogr., 229 (1982) 57-65.

51   C.E. Dunlap, S. Gentleman and L.I. Lowney, J. Chromatogr., 160 (1978) 191-198.

52   J.L. Meek, Proc. Natl. Acad. Sci. USA, 77 (1980) 1632-1636.

53   M.J. O'Hare and E.C. Nice, J. Chromatogr., 171 (1979) 209-226.

54   H. Nau and K. Biemann, Anal. Biochem., 73 (1976) 139-153.

Chapter 5
ANALYTICAL MEASUREMENTS OF ENDOGENOUS PEPTIDES

5.1 INTRODUCTION

The objective of quantitative analysis in general is to objectively measure the amount of a selected endogenous compound in the presence of many other compounds, where the concentration of some of those other compounds may greatly exceed the concentration of the target compound, and to perform that analysis in a fast, facile, and objective manner with a maximum of three crucial experimental parameters- accuracy, precision, and molecular specificity. Molecular specificity is the most critical factor, from the viewpoint of this book.   Furthermore, in a biologic system, it is occasionally necessary to obtain the metabolic profile of a selected class of compounds (for example, neuropeptides, steroids, fatty acids, amino acids, drugs, organic acids, or other compound) in a biologic extract, and to observe the quantitative or semi-quantitative metabolic relationships among the constituents in that family of compounds before, during, and after a physiological experiment. Even though these analytical goals can be stated easily, they are in most cases rather difficult to experimentally achieve. These difficulties notwithstanding, however, accurate analytical measurements must be made of individual compounds to enable a rational understanding of cellular, neuronal, and clinical events. Workers in the diverse fields of biology, biochemistry, clinical chemistry, and analytical research are continuously developing appropriate state-of-the-art analytical methodology to be able to experimentally achieve those goals.   Each selected analytical methodology possesses its individual limits and advantages, and these comparative aspects will be discussed in this chapter and illustrated with experimental data.

Three of the more frequently-used assay methods will be discussed.   RIA, BA, and RRA techniques will be reviewed. Two other techniques, LC and MS, are also used.   LC is discussed in greater depth in Chapter 4, MS in Chapter 6, and MS analytical measurement methods in Chapter 7. This chapter presents several topics including the basic principles, an interpretation of the crucial aspect of molecular specificity, representative analytical data, and the inherent advantages and disadvantages of RIA, BA, and RRA.

5.2 RADIOIMMUNOASSAY OF PEPTIDES

5.2.1 Basic Principles

RIA is a well-suited and broadly applicable method for rapid, sensitive, convenient, and relatively inexpensive analysis of biologically important compounds (1). In an RIA experiment, the peptide to be measured (or a synthetic conjugate) acts as an antigen and is injected into an animal where the animal's immune system is activated to produce an antibody to that injected antigen. Those antibodies are extracted, purified if necessary, and reconstituted in an appropriate RIA buffer. The RIA method was developed originally for study of insulin (1-4). Immunoassays are based on the binding specificity of an antibody for its (putatively unique) antigen and may be used analytically to measure either the endogenous antigen or antibody. A radioactive isotope generally labels the antigen, and the term "radioimmunoassay" is used. The measurement depends on the ability of the unlabeled antigen (Ag, the unknown) to inhibit or compete with the binding of the radioactive antigen (Ag*) by an antibody (Ab). This competition is schematically represented in the author's conceptualization (Figures 5.1 and 5.2) of the two experiments which are performed for RIA. In experiment 1 (Figure 5.1), a known amount of Ag* is incubated with Ab. The Ab molecule has a fixed number of sites which are available to bind the Ag molecules. The number of sites may range from one to several. Following the incubation, the remaining unbound Ag* molecules are in the presence of the Ab-Ag* complex, the latter is precipitated, and the amount of remaining radioactivity is measured. That remaining radioactivity in turn relates directly to the remaining amount of Ag*. In the second experiment in RIA (Figure 5.2), a known amount of Ag* is mixed with the unknown amount of Ag, where the latter represents that molecule which is the objective of the analytical measurement. This mixture of antigens is incubated with another portion of Ab which still contains the same fixed number of binding sites as in experiment 1, but these sites now competitively bind both the Ag and Ag* antigen molecules. After incubation, a mixture of Ag and Ag* antigens plus the Ab-Ag-Ag* complex is present, in which the Ag and Ag* molecules have competitively bound to the fixed number of binding sites on Ab. Again, the Ab-Ag-Ag* complex is precipitated, and the amount of radioactivity which remains is due to the remaining unbound Ag*. The amount of endogenous peptide is determined from a standard RIA curve (See Fig 5.3) which plots the ratio of bound/free Ag versus the amount of unlabeled Ag.

Femtomoles of a compound can be analyzed by RIA (4). Whenever the size of the original Ag molecule to be analyzed is too small (less than 4,000

EXPERIMENT 1

Ag* Ag* Ag*
Ag* Ag* Ag*
Ag* Ag* Ag*
Ag* Ag* Ag*

Known amount of
radioactive–labeled
antigen, Ag*.

Ab

Antibody molecule, Ab, with a
fixed number of sites to bind
antigen molecules.

Ag*
   Ag*
Ag*

Remaining
unbound Ag*.

Ab

Antibody–Antigen*Complex
Ab · Ag*

1. Precipitate Ab·Ag*
   complex
2. Measure amount of
   remaining radioactivity
   which is due to
   remaining unbound Ag*.

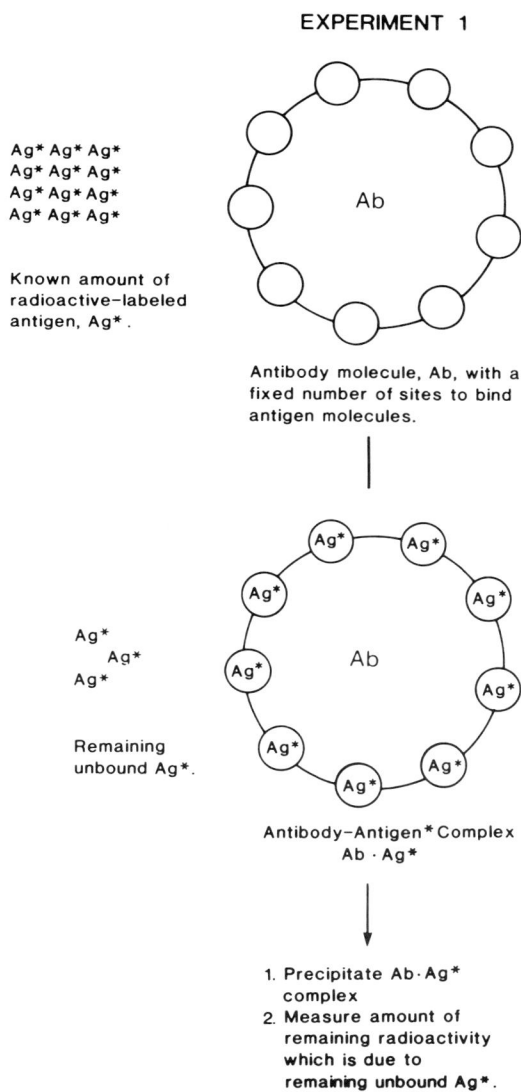

Fig.5.1. Schematic representation of radioimmunoassay – Experiment 1.

EXPERIMENT 2

Ag⁰ Ag* Ag⁰ Ag⁰ Ag* Ag⁰
Ag⁰ Ag* Ag⁰ Ag⁰ Ag* Ag⁰
Ag⁰ Ag⁰ Ag⁰ Ag* Ag⁰ Ag⁰
Ag* Ag⁰ Ag⁰ Ag⁰ Ag* Ag⁰
Ag⁰ Ag* Ag* Ag⁰ Ag* Ag⁰
Ag⁰ Ag⁰ Ag⁰ Ag* Ag⁰ Ag*

Known amount of radioactive-
labeled antigen, Ag*, plus an
unknown amount of unlabeled
antigen, Ag⁰.

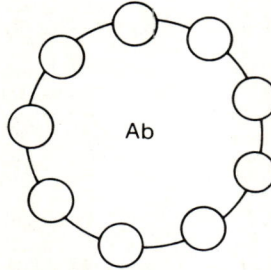

Antibody molecule, Ab, with a
fixed number of sites to competitively
bind unlabeled, Ag, and radioactive-
labeled, Ag*, molecules.

Ag⁰ Ag⁰ Ag*
Ag⁰ Ag* Ag⁰ Ag⁰ Ag* Ag⁰
Ag⁰ Ag⁰   Ag* Ag⁰
      Ag⁰ Ag⁰ Ag⁰ Ag* Ag⁰
Ag⁰ Ag* Ag* Ag⁰ Ag* Ag⁰
Ag⁰   Ag* Ag⁰

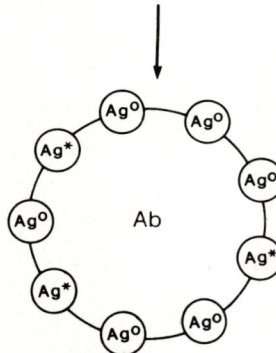

Antibody-mixed antigen complex Ab·Ag⁰·Ag*
in which unlabeled, Ag⁰, and radioactive-
labeled, Ag*, antigen have competed to bind
to the fixed number of binding sites.

1. Precipitate Ab Ag⁰·Ag* complex
2. Measure amount of remaining
   radioactivity which is due to
   remaining unbound Ag*.

Fig. 5.2.   Schematic representation of radioimmunoassay - Experiment 2.

TABLE 5.1. Validation procedures for immunoassays.

1. Parallel studies can be performed in another assay system which is based on a completely different principle of measurement (enzymes, mass spectrometry) or at least utilize preliminary chromatographic separation.
2. Use of internal standards.
3. Enzymatic destruction of the antigen.
4. Slope comparison of the Ag dose-inhibition curves (unknown sample versus standard).
5. Screening studies with substances known to be present in the samples that might theoretically cross-react and therefore inhibit immunoassay.
6. Co-migration of immunoreactivity and tissue samples with antigen.
7. Assays under conditions that markedly alter the values in a positive or negative fashion. (An example of an unexpected reaction in a routine screening is a positive urine immunoassay for morphine in individuals ingesting bread or coffeecake that contains poppy seeds).
8. Analysis during either pregnancy or pathological states where binding proteins in plasma might affect the assay.
9. Different laboratories performing analysis on the same samples.
10. Sequential analysis of samples which have been stored over a period of time to determine how decomposition may affect the assay.

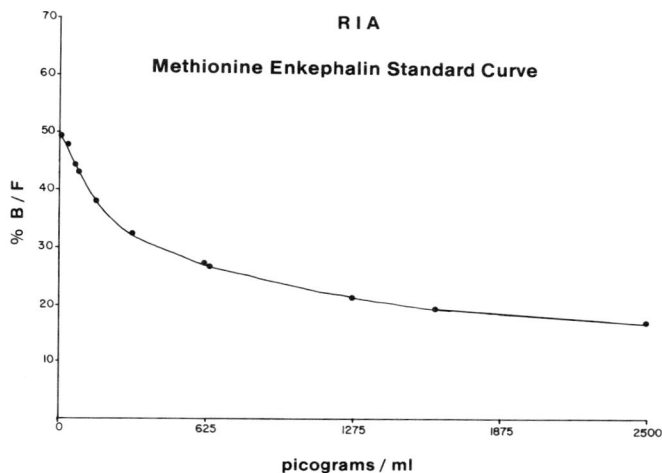

Fig.5.3. Calibration curve constructed for RIA of ME.

daltons), it can be chemically conjugated to a larger carrier molecule such as ovalbumin, thyroglobulin, or bovine serum albumin to increase its immunogenicity. For example, up to 30 molecules of the ME are chemically conjugated to ovalbumin (5; see Fig. 5.7 below). The complex of small peptide:large carrier protein is injected into an animal and antibodies are raised to the antigenic complex, but, it must be realized, not necessarily only to the original oligopeptide.

Several pertinent radioisotopes are available for labeling a protein and have half-lives ranging from eight days (^{131}I), 56 days (^{125}I), 12.3 years (^{3}H), and 5,730 years (^{14}C). The popular usage of ^{125}I and ^{131}I as radioindicator atoms for biological molecules derives from these differences in respective radioactive decay rates. Chloramine T is a chemical method generally used to incorporate an iodine atom into either a Tyr amino acid residue or, to a lesser extent, the other aromatic amino acid residues. This chemical incorporation method has advantages of reproducibility, rapidity, and efficiency of iodination in the absence of carrier. It has been observed that oxidation of several sensitive amino acid residues may occur during iodination. In a typical two-stage labeling method, the low molecular weight carrier is iodinated in the absence of the protein and combined with the carrier protein via a stable chemical bond. It is important to consider the fact that high levels of iodination, which may be theoretically possible, and initially even thought to be desirable for increased sensitivity, may result in radiation damage and cause antigen denaturation, decreased immunoreactivity stability, and increased immunoassay variability. Several methods are available to separate free from bound Ag* and include a second Ab, salt precipitation, adsorption onto charcoal, or other nonspecific adsorbant or two-phase systems.

No matter which RIA system is used, it is of paramount importance to fully evaluate the molecular specificity and detection sensitivity of the system in order to know what level of confidence to place on the resulting analytical data. Rigorous experimental controls are needed for this type of assay system, because it cannot be assumed that immunoassays which give sensitive and reproducible results in a pure control buffer will necessarily give valid results for measurement of the target endogenous molecule in the biologic matrix or an appropriate extract. While the Ag and/or Ab may be heterogeneous, some proteins, polypeptide hormones, and enzymes are known to be secreted in multiple molecular precursor forms (see discussion in Chapter 2) and may exhibit partial structural homology with other hormones and enzymes, undergo partial proteolysis, or become attached to naturally-occurring inhibitors or carriers that may or may not affect immunologic reactivity. For all of these reasons, numerous procedures have been developed to validate immunoassay results, and these procedures are tabulated in Table 5.1.

5.2.2 Measurement of Peptides with Radioimmunoassay

The regional distribution of a selected peptide, ACTH, in brain has been critically analyzed (6). The hypothesis of the authors is that the pituitary is the sole site of synthesis of ACTH and that it occurs in other intracranial sites for reasons other than synthesis. In the smaller rat brain, ACTH and endorphins have been found to be widely distributed, but that wide distribution probably is due to the small physical dimensions of that brain. ACTH and the thyrotropin stimulating hormone, TSH, while widely distributed in a rat brain, are found only in the hypothalamic region of the primate brain. Peptide transfer from the pituitary to the brain may occur by several other anatomic routes including the stalk, CSF, or a neural route. The commercially available hypophysectomized brain tissue used in many previous studies may have residual pituitary fragments present in that surgical preparation. In addition, artifacts may also be introduced by the extraction and/or assay system itself. Inadvertent degradation of the labeled Ag* and/or Ab may also result in apparent decreased binding. It was noted that the use of antisera raised against purified pituitary ACTH may provide better antisera to the contaminants than to the ACTH itself. Discrepancies were noted between intensity of immunohistochemical localization in tissue and the immuno-chemical content of a peptide. For example, the tissue level of insulin was found to be 25 times higher than that of plasma. However, utilizing ODS cartridges for preliminary chromatographic purification before RIA, the normal one ng g^{-1} level was found in guinea pig tissues. It was concluded that accidental contamination with a porcine-like insulin may have occurred in previous analyses.

RIA is performed conveniently utilizing commercially available kits. One set of experiments will be described here in detail to illustrate the RIA procedure. A standard of 2,500 pg ME is sequentially diluted by factors of two to produce a series of standard solutions ranging down to 39 pg ml^{-1}. This series of batch-wise diluted standard ME solutions is used to construct a calibration curve (Figure 5.3). Because the concentration of immunoreactive enkephalin in a biologic sample may not be known, it is necessary to study both undiluted and two dilutions (100x and 10,000x) during initial experiments. Once the unknown concentration of immunoreactive ME (pg ml^{-1}) is determined from the standard curve, the following equation is utilized to calculate the concentration of endogenous immunoreactive ME [ir-ME]:

$$[\text{ir-ME}] = \text{pg ml}^{-1} \times 0.5 \text{ ml} \times 10 \text{ g}^{-1} \text{ wet wt. of tissue} \qquad (1)$$

RP-HPLC data (isocratic mode) for the peptide-rich fraction extracted from the first tooth removed from each of three human patients are shown in Figures 5.4, 5.5, and 5.6 to demonstrate several experimental features including resolution, "cleanliness", and also inter-patient biological variability of the individual peptides contained in the peptide-rich fraction. The chromatograms of only the first tooth removed are shown in those figures for these three patients because these samples contain the highest concentrations of ir-ME. A standard solution of synthetic ME in the RP-HPLC isocratic mode elutes at 5.2 min, a retention time noted by the arrow at that time in the Figures. Therefore, the fraction eluting at this retention time is collected manually in all three chromatograms. The rightmost arrow denotes the elution time of synthetic LE.

In the first case, Figure 5.4 (patient DD), immunoreactivity is also found at 5.9 min in addition to the peak at 5.2 min. Further characterization of this peak at 5.9 min awaits both RRA assay analysis and MS structural characterization. While it is possible that facile production of ME sulfoxide may occur by these experimental manipulations, conditions conducive to this oxidation are rigorously avoided. There are reports of an endogenous sulfoxide reductase enzyme which, if present in tooth pulp, may exert a metabolic control point during ME and precursor metabolism. (7).

Figure 5.3 contains the RIA standard curve obtained from the series of diluted synthetic ME solutions. This curve extends from a bound-to-free (B/F) ratio of 50%, down to a ratio of 17%. These two B/F ratios correspond to a range of 2,500 pg ml^{-1} to 30 pg ml^{-1} of synthetic ME. Values of ir-ME of the RP-HPLC peaks collected at the ME retention time from the first teeth extracted from each patient are noted on the calibration curve. Table 5.2 collects both the ir-ME pg ml^{-1} from the curve and ng g^{-1} values of all teeth calculated from equation (1). (This table also collects RRA data of the same tissue. This RRA data is discussed later). In addition to data corresponding to the three human patients mentioned above, Table 5.2 also contains RIA and RRA measurements for four additional patients (CB, TH, SG, and SS). While CB is also a control patient, the other three are treated patients. Those three treated patients had a premolar tooth in each maxillary quadrant banded and connected with a spring to apply a strong orthodontic force 30-45 min prior to the removal of the two maxillary

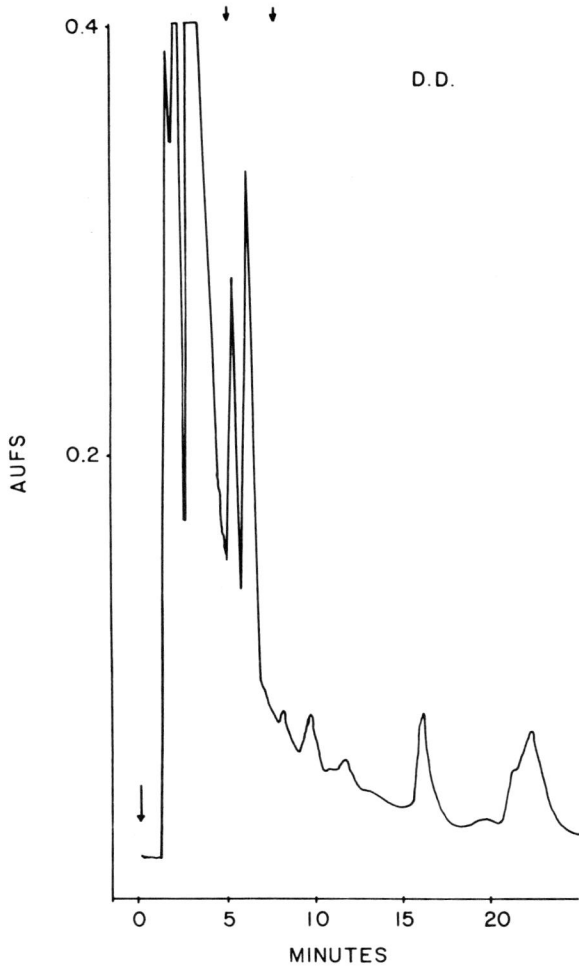

Fig. 5.4. RP-HPLC of peptide-rich fraction extraction from tooth pulp of patient D.D. Arrows indicate retention time for ME and LE, respectively (12).

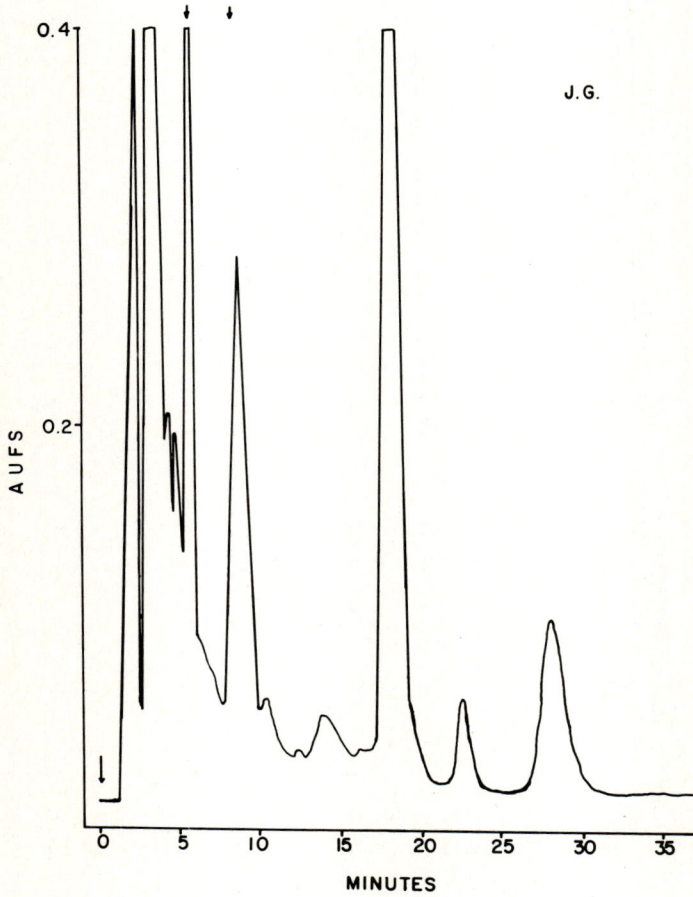

Fig. 5.5. RP-HPLC of peptide-rich fraction extracted from tooth pulp of patient J.G. Arrows indicate retention time for ME and LE, respectively (12).

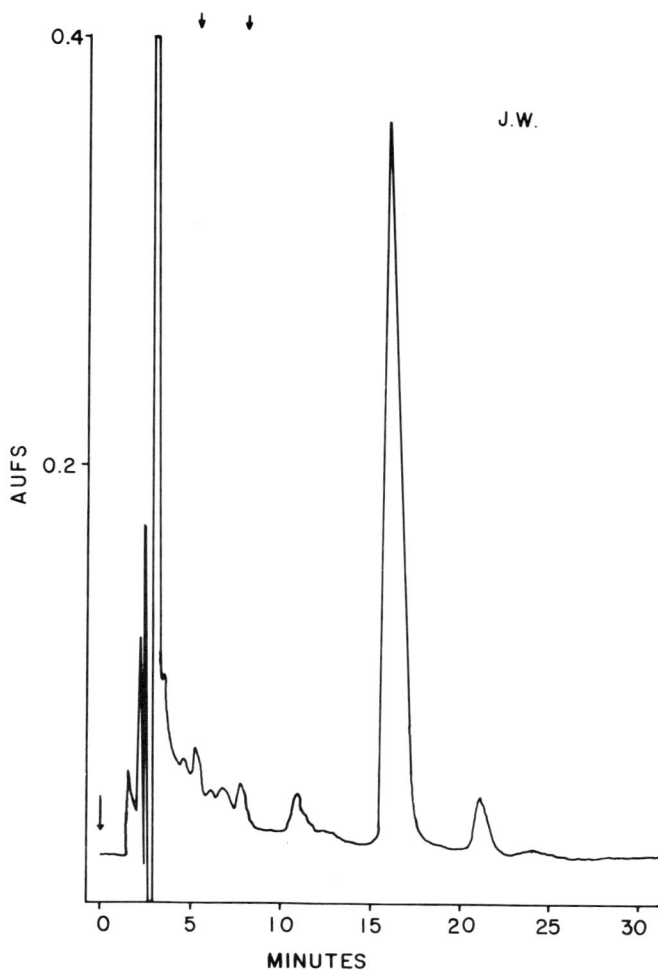

Fig. 5.6. RP-HPLC of peptide-rich fraction extracted from tooth pulp of patient J.W. Arrows indicate retention times of ME and LE, respectively (12).

TABLE 5.2. Opioid concentrations measured as ng enkephalin equivalents g^{-1} wet wt. tissue in a radioreceptor assay based on displacement of ^{3}H-dihydromorphine and in a radioimmunoassay (12).

PATIENT	TOOTH NO.	WET WEIGHT (g)	RIA	RRA
A. CONTROLS				
DD	5*	.1181	68	**
	12	.0576	4.9	521
	21	----	---	---
	28	.1181	5	47
JG	1	.0394	15.9	6.726
	16*	.0252	500	3,155
	17	.0543	7.2	5,608
	32	.0649	16.6	1,495
JW	--*	.0123	1,138	6,098
	--	.0273	7.1	421
	--	.0478	1.3	1,339
CB (a)	5	.0710	176	292
	12*	.0843	148	1,751
	21	.0496	51	107
	28	.0556	155	79
CB (b)	5	.0710	38	1,215
	12*	.0843	14	867
	21	.0496	5	968
	28	.0556	12	809
B. TREATED				
TH (a)	5*	.0448	19	91
	12	.0456	4	52
	21	.0396	n.d.	67
	28	.0402	15	74
TH (b)	5*	.0448	n.d.	81
	12	.0456	4	86
	21	.0396	n.d.	83
	28	.0402	40	114
SG (a)	5*	.0486	12	n.d.
	12	.0489	33	22
	21	.1107	17	29
	28	.0925	1	n.d.
SG (b)	5*	.0486	4	66
	12	.0489	16	21
	21	.1107	6	25
	28	.0925	1	n.d.
SS (a)	5*	.1303	32	68
	12	.1056	2	28
	21	.0773	5	29
	28	.0732	22	80
SS (b)	5*	.1303	2	39
	12	.1056	2	19
	21	.0773	3	28
	28	.0732	n.d.	23

(a) HPLC fraction collected at synthetic ME (YGGFM) retention time.
(b) HPLC fraction collected at synthetic LE (YGGFL) retention time.
*first tooth extracted
**sample used for MS corroboration
n.d. = not detected
Patients TH, SG, and SS were fitted with springs.

premolar teeth and the two mandibular premolar teeth. The purpose of this orthodontic treatment is to analytically observe any changes in the RIA and RRA concentrations of enkephalins in tooth pulp as a function of the amount of applied stress (8).

It is significant to note that no obvious correlation exists between the RP-HPLC peak height (Figures 5.4, 5.5, and 5.6.) and the RIA values of ng enkephalin g^{-1} wet weight tooth pulp tissue (Table 5.2). Furthermore, no correlation is observed between ng g^{-1} values and the patient's age, tissue weight, and sex. The amounts of endogenous ir-ME measured in this study compare to the first measurement of ME in tooth pulp by MS methods, where microgram amounts were found (9). It will be noted that, except for patient CB, who received additional intravenous medications (valium and stadol), the three other controls demonstrate a distinct pattern, where the first tooth removed contains a large quantity of ir-ME. Subsequent teeth removed had orders-of-magnitude lower quantities of RIA-ME. On the other hand, data corresponding to the three treated patients (TH, SG, and SS) indicate that generally, lower amounts of endogenous ir-ME are found in all teeth. The comparative amounts of the two enkephalin levels are not incompatible with the hypothesis that stress will mobilize the enkephalin-producing and -metabolizing processes, both of which will lead to a diminished level of ME when that tissue is under stress.

The three measurements TH(b), SG(b), and SS(b) correspond to the use of the commercial ME Ab to measure that RP-HPLC fraction where LE elutes. The purpose of that measurement is twofold: determine the extent of cross-reactivity of the Ab and determine the presence of material that will elute at the retention time of LE, is not necessarily LE, but interacts with an ME Ab. Those data indicate that low picogram amounts of such material is present in that fraction.

Great care is taken to ensure that a "snapshot" of the metabolism of endogenous enkephalin is obtained. The tissue is acquired rapidly, tissue is placed immediately into liquid nitrogen to inhibit chemical and or enzymatic interconversion, acetic acid is used to precipitate enzymes and to preserve smaller peptides present in this tissue extract, and the RIA antibody is relatively specific. Furthermore, fast atom bombardment mass spectrometry (FAB-MS, 10) buttresses the fact that this extracted peptide is ME, as attested to by production of the appropriate protonated molecular ion $(M+H)^{+}$ at mass 574 in the analysis of one of the extracts (Patient DD; Tooth number 5).

It is interesting to note that the RP-HPLC chromatogram of one patient (DD) has another large peak eluting at 5.9 min. following the synthetic ME retention time (5.2 min.). Due to results obtained with RP-HPLC of synthetic peptide standards which are substantiated by MS, this second peak at 5.9 min. is not ME. However, it is significant to note that this RP-HPLC fraction does cross-react with the commercial Ab raised against ME and underlines the importance of using high resolution RP-HPLC before analysis in an effort to at least minimize, but not completely avoid, ambiguity. Furthermore, this observed cross-reaction raises doubts about the putative molecular specificity of that Ab.

5.2.3 Molecular Specificity of Radioimmunoassay Method

The molecular specificity of the RIA analytical method is putatively quite high. However, recent research indicates a growing concern and acceptance that, even though immunological criteria of identity may have been adequately satisfied, these criteria do not constitute rigid, objective, nor unambiguous structural identification and certainly do not constitute structure elucidation of an endogenous compound. Determinants of the active site(s) on an Ab molecule are not exclusive to only one natural Ag. This factor represents a crucial and pivotal experimental and conceptual fact which constitutes a critical limitation of the RIA method. This limitation must be remembered, before, during, and after any RIA analytical measurement because that fact makes it logically impossible to exclude all other candidates for the endogenous ligand in an immunochemical experiment, even if HPLC precedes RIA measurement. Clearly, an independent standard analytical method is needed to calibrate any RIA method (see Table 5.1).

RIA is probably the one analytical method which is most used around the world for quantification of endogenous peptides such as somatostatin, insulin, enkephalins (11), endorphins, and many other peptides of interest in biological tissue extracts. The speed of the measurement process is high, sensitivity attains the femtomole level, and a large number of analyses is run conveniently. If a small peptide molecule is the objective of that measurement, then it is imperative to determine how and where that small peptide molecule is chemically conjugated to the larger carrier molecule because the Ab which is immunogenically raised to the peptide:protein Ag complex can react only towards that portion of the molecule which is free and available for interaction with binding sites on the Ab surface. That available structure may be the free N- or C-terminus, the middle portion of the molecule, the entire molecule, an internal hydrophobic volume, or whatever three-dimensional convoluted molecular surface is presented to and interacts with an active

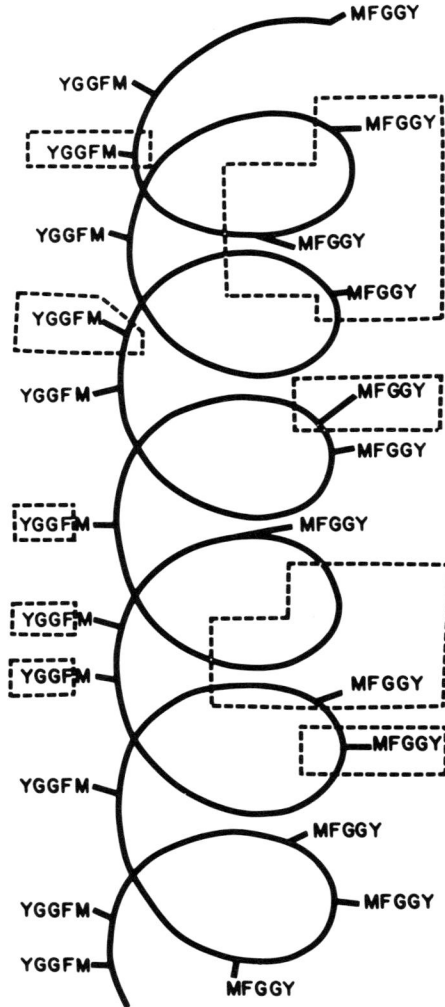

Fig.5.7. Schematic representation of an immunogenic conjugate of 23 ME molecules per molecule of thyroglobulin.

site of the Ab. This multifaceted binding phenomenon is schematically illustrated in the drawing in Figure 5.7 which demonstrates the author's concept of the 23 ME molecules which are chemically bound through the pentapeptide's carboxy terminus to one thyroglobulin protein molecule (5). This peptide:protein complex is injected into an animal whose immune system then raises an Ab (or antibodies) in response to the presence of this antigenic complex. The Ab may be formed in response to immunological interaction(s) with one or more of the structurally different sites on the peptide:protein complex. Those possible sites of interaction are represented by the dashed-line boxes. In some cases, only the N-terminus of the pentapeptide ME molecule might produce an Ab while, in other cases, interaction may involve several different peptide: protein regions.

While the foregoing concepts attempt to demonstrate the crucial fact that it is imperative to always unambiguously determine the molecular specificity of any assay method, that goal is virtually impossible to attain with most analytical methods, especially when working with a complex biological mixture. However, an increasing number of workers are at least using HPLC separation of a biologic mixture of peptides before RIA (11, 12) or RRA (13) of one selected peptide. While that combination of analytical methods will greatly improve the molecular specificity of the analysis, that combination still cannot rigidly guarantee unambiguous compound identity.

Table 5.3 lists one of the first sets of RIA measurements of the three neuropeptides ME, LE, and beta-endorphin in selected rat brain regions where ODS RP low resolution mini-columns were used for rapid preliminary chromatographic purification of peptides (11). This effective, but still limited purification procedure is predicated on the fact that subsequent RIA will be performed with highly specific antibodies. A Sep-Pak is utilized where the C-18 RP hydrophobic surface is activated first by washing with acetone (2 ml) and distilled water (5 ml). Following application of the peptide sample, the Sep-Pak cartridge is washed twice with acetic acid (4%, 2ml) to remove water-soluble salts and compounds. The peptides (enkephalins and endorphins) were eluted with hydrochloric acid: acetone (25:75, 1.5 ml, 0.2 M). Recovery of the peptides is high (96-98%) and is unaffected by brain tissue concentration. For the data listed in Table 5.3, the beta-endorphin antibody was directed towards the C-terminus of the peptide and cross-reacts 100% on a molar basis with beta-LPH. While the original data was presented as pmol enkephalin-equivalents immunoreactivity mg^{-1} protein, the data is recalculated in terms of ng peptide g^{-1} tissue, factoring in the 10% of protein in brain tissue (14).

TABLE 5.3. Peptide concentrations in rat brain by radioimmunoassay
(ng peptide g^{-1} brain) (11).

BRAIN AREA	ME	LE	β-END
Caudate	764	103	-
Hypothalamus	471	134	208
Septum	210	34	20
Brain stem	143	34	-
Mid brain	117	29	10
Cortex	101	14	-
Hippocampus	63	11	-
Cerebellum	55	7	-

The variability of the ratio of ME/LE in these data reflects differential excision processing of pentapeptides from the much larger precursor molecules, a mechanism which is discussed in Chapter 2.

The presence of ME-like immunoreactivity was demonstrated in a population of olivocochlear fibers and in efferent terminals of inner and outer hair cells in guinea pig cochlea (15). It could not be determined immunocytochemically whether the sets of enkephalinergic and cholinergic fibers in the cochlea completely coincided. Again, in the hope of not being redundant, but rather to underline an important theme of this book, it was stated by these authors that caution is demanded in interpreting immunocytochemical data because of the limited chemical specificity which is inherent in immunological techniques. The amino acid homologies and consequent antigenic similarities among the many endogenous enkephalins and endorphins, which can extend even to functionally unrelated proteins (See Chapter 2), demand verification of the presence of endogenous substances of interest by using complementary chemical and physical techniques of identification and careful characterization of the antisera. Towards that end, this study utilizes HPLC of cochlear sonicates coupled with RIA of ME sulfoxide. The methionine residue in methionine enkephalin was oxidized quantitatively (30% H_2O_2, 30 min., room temperature) to methionine sulfoxide. It was noted that the maximum sensitivity for optical detection of peptide synthetic standards during HPLC was two to five ng (16).

Those authors go on to state that satisfaction of immunological criteria of structural identity is not sufficient for identification of actual endogenous substances (15). This limitation derives from the fact that the determinants of the active site of the Ab are not exclusive to only one natural Ag, making it logically impossible to exclude all other candidates from the endogenous ligands in an immunochemical experiment. Physical means must be utilized to demonstrate clearly that the proposed endogenous ligands to the antiserum used in immunocytochemistry are actually present in that tissue of interest. It is clearly possible (and it has been done in several cases) to inadvertently raise an antiserum (also, or even instead of) against a minor impurity in a synthetic Ag which was used to immunize the host animal. At the low dilutions of antisera often used in histochemistry experiments, it is possible that false positives may easily obscure biologically significant results. The use of an amount of Ag in absorption controls which is just sufficient to absorb out the Ab of interest will also serve to guard against this last possibility.

This study (15) demonstrates the developing concept that the presence of peptides in sensory systems is becoming the biological rule rather than the exception. Peptide biochemistry possesses a high level of heterogeneity which makes neuroactive peptides particularly well-suited to mediate the interface between organisms and their environments, and to provide a means of peripheral acuity to match the unique requirements of individual organisms (See Figure 2.1). The increasing abundance of data relating to localization of peptides in the peripheral sensory structures furthermore emphasizes the capacity for sophisticated biological data processing which peripheral sensory neurons possess.

Various peptides found in brain have been found to also occur in the neural retina (17). The major form of CCK in the brain is generally the sulfated octapeptide. HPLC followed by RIA of CCK peptides was performed in a study of the vertebrate neural retina. A Sep-Pak C_{18} cartridge was utilized before a C_{18} RP-HPLC analytic column, where the latter was eluted isocratically with acetonitrile: TEAP buffer (22:78; pH 6.5) The octapeptide CCK sulfate is the primary form of CCK in the vertebrate retina, at least in the rat, bullfrog, and cow.

5.2.4 <u>Other neuropeptides studied by combining high performance liquid chromatography separation and radioimmunoassay measurements</u>

Corticotropin-related peptides were measured in extracts from the intermediate lobe of the rodent pituitary gland with a combination of HPLC and RIA (18). Electrostimulation (15 Hz, 5 ms, 10 V, 15 min) of the mandibular nerve influences the concentration of substance P-like immunoreactivity obtained in the superfusate of dental pulp (19). The undecapeptide substance P may play a role in the development of the inflammatory response associated with both injury and pain in peripheral tissues (see chapter 2). The latter types of studies are crucial in experiments designed to elucidate the molecular basis of nociceptive processes.

It is worthwhile and constructive to remember that the question of optimal molecular specificity of the analytical measurement of a biologic compound is not limited to peptides. Other workers have noted that apparently both the sensitivity and molecular specificity of RIA measurement of prostaglandins can not lead, but rather must follow in a parallel fashion, those developments in the field of combined GC-MS, an instrumental method that offers optimal molecular specificity (20).

5.3 BIOASSAY

The objective of a BA method is to provide an analytical system which simulates as much as possible the in vivo biologic system which contains receptors which bind specific molecules. For example, the vas deferens or guinea pig ileum tissues are frequently utilized for BA of enkephalins.

In a typical BA experiment such as muscle contraction, an appropriate length of the tissue is supported on a strain gauge, and the system is bathed with a suitable medium. After application of a synthetic opioid alkaloid, peptide, or HPLC-purified fraction to the solution bathing the strip of tissue, the resulting tissue contraction is measured and the amount of contraction is compared to a calibration curve.

The advantage of a BA is that the analysis most closely approximates the response of a living organism. On the other hand, the limitation inherent in BA analysis is that other interfering substances are often present in an extract, even in an HPLC-purified fraction, which may also evoke the same biological response.

While all of the ramifications of BA techniques, methodology, and resulting data will not be fully elaborated in this book, it may be clear to the reader that from a simplistic viewpoint, BA most closely approximates the RRA method. This similarity derives from the fact that an endogenous compound first binds to its putatively unique receptor on the tissue used for BA to evoke a particular biological response (see Figure 2.1).

5.4 RADIORECEPTOR ANALYSIS

5.4.1 Background

A receptor is defined as that macromolecular complex which is on the surface of a cell and which will specifically interact with and bind a selected molecule. Putatively, this interaction is structurally unique. The macromolecular complex may be comprised of protein, lipid, and saccharide portions. The three-dimensional aspects of the receptor are crucial to interact with a particular ligand molecular conformation. Various ligands have different experimental binding characteristics which may be represented by their binding avidity and concentration range.

The following limited discussion will place into the perspective of this book the history of receptor analysis and its molecular specificity.

When performing the type of biological research described in this book which focuses on the selected specific field of opioid peptide neurochemistry, it is important to have an assay method which possesses a broad range of molecular specificity so that that assay can be used to screen for the presence of all opioid receptor-binding compounds in either a biological mixture or all of the fractions separated by RP-HPLC.

The theme developed in this book is that MS provides maximum molecular specificity to quantify specific endogenous peptides in a biologic matrix. RIA methods display one of the highest levels of molecular specificity compared to the other assay methods, but does not have the highest level of molecular specificity which is available only with the MS method described in Chapter 7. Nonetheless, an RIA system of high molecular specificity can be used for measuring peptides or, on the other hand, an RIA with a broad-based low specificity may be utilized for screening chromatographically-separated opioid peptide fractions from a biologic matrix. However, in the latter case of screening, specificity is still generally too high with RIA.

An RRA method is the clear method of choice when a very broad response and lower level of molecular specificity is needed to rapidly and efficiently, with a high level of sensitivity, screen chromatographic fractions for opioid receptor activity. The purpose of this section is to discuss the basic principles of the RRA method.

The RRA type of measurement is based on the experimental observation that opiate-receptor interactions occur with specific endogenous brain receptors, a fact central to the in vitro studies discussed in Chapter 2 which showed the presence of brain enkephalins. Figure 2.3 shows a representation of the polypeptide macromolecular opioid receptor complex which is thought to be located on a synapse. Preliminary receptor studies demonstrate that some binding does occur to brain membranes, even though that binding is non-specific and only 2% is stereospecific. Later, opiate ligands with a high specific activity of radioactivity made it possible to examine opioid receptors which have a high affinity for the opiates and also permitted the use of low ligand concentrations, an experimental feature which significantly decreased the amount of non-specific binding (21).

Rapid separation and washing of the membrane reduced non-specific entrapment of radioactive ligands. Up to 90% stereospecific binding was demonstrated by using these experimental protocols. Pharmacological data indicated that the binding of the opioid receptor population was saturable (a phenomenon that demonstrates a finite number of receptors) and reversible, with both association and dissociation constants independent upon the ligand being used. Bacitracin is used to inhibit enkephalin destruction (22).

It is clear from the foregoing brief review that a range of measurement methods and molecular specificities is available to the peptide chemist. The maximum obtainable specificity is provided by MS techniques while RRA provides a high but, vis-a-vis MS, lower molecular specificity. That known limitation notwithstanding however, many electrophysiology experiments require a detection system with a high sensitivity, but very broad and low level of molecular specificity to provide a fast and facile screening method for testing multiple HPLC fractions for opioid receptor activity. The RRA method fulfills the latter criteria.

Initial studies indicate that interaction between an endogenous brain fraction and the opioid receptor in tissue is reversible and competitive (23). Human CSF and brain extracts were tested with a receptor (24, 25) and different receptor selectivities were found with various opioid enkephalins (26). Pain perception and endorphin levels in CSF (27) and intercranial hydrodynamic dysfunction in brain were studied (28). $Dynorphin_{1-17}$ in the hippocampus (29) and endorphin levels in CSF during alcohol intoxication and withdrawal were studied (30). Electrophoretically-separable endorphins in human CSF were characterized, where procedures were guided by a specific RRA (31). Inhibitors of the narcotic receptor binding in brain extracts and CSF were studied (23). Brain-pituitary opioid mechanisms and specific influences and interactions of the opioid receptor binding were studied (22).

A variety of other aspects of the opioid receptors was studied and included stereospecific binding (32), multiplicity of opioid receptors (33), receptor gradients and monkey cerebral cortex (34), enkephalin binding to opioid receptors by zinc ions and the possible physiological importance of that ionic interaction in the brain (35), displacement reactions employing heterologous tracer ligands, a review (36), and a detailed review of the molecular properties of opioid receptors (37).

It has been shown by using a variety of immunohistochemical techniques that structures within the limbic system of the brain contain some of the highest concentrations of opioid receptors (38). The limbic system is a loosely-defined system generally considered to contain the cingulate gyrus, parahippocampal gyrus, hippocampus, septum, amygdaloid body, hypothalamus, and the olfactory bulb (39; see Figure 3.2).

Following an assay of opioid peptides by inhibiting tritiated naloxone and tritiated LE binding in brain homogenates, it was concluded that the opioid peptidergic system has several agonists possessing different characteristics which interact with more than one type of receptor (40). The hydrophobic moiety of the benzene ring of the enkephalin molecule may be one of the major structural factors which is operative in differentiating binding sites (41). These multiple opiate receptors were measured in different brain regions and the distribution and differential bindings of opiates and opioid peptides were studied (42). A review of the binding, molecular, and structural parameters are reviewed (43).

Based on the assumption that the mu receptor is the mediator of analgesia, two distinct mu and delta receptors are located in a "opioid receptor complex" in which the two receptors interact with each other (44). The occupation of one receptor by its specific ligand is hypothesized to induce a conformational change in the second receptor, which in turn allows the coupling of that receptor to its effector. Ethanol inhibits opiate receptor binding (45). Endogenous ligands for the kappa opiate receptors have been studied (46).

The distribution and physiological significance of opiate receptors in the brain is outlined by a discussion of receptors, autoradiographic localization of opioid receptors, and opioid receptors associated with the sensory, limbic, and neuroendocrine systems (38). The different types of opioid receptors are classified (47). It is stated that a receptor consists of both a recognition site to which the ligand binds, and an effector region that translates that binding into biochemical events which lead to a biologic response (see Figure 2.1). The biochemical characterization of receptors is discussed and includes competitive displacement assays, saturation analysis, selective protection assays, and distribution. It is noted that binding assays provide information only by interaction of a drug with a recognition site, and that those binding results must be supplemented by a pharmacological analysis of those effects.

As a means to illustrate in more detail the RRA methodology, data are presented in the next section in an effort to describe for the first time measurement of endogenous LE and ME in human tooth pulp tissue extracts with an RRA (13). Tooth pulp tissue is chosen as the model tissue for this type of study of nociceptive processes because it has been hypothesized (48) that the only sensory output from tooth pulp tissue is pain. Advantage is taken of the increased level of the opioid receptors in the limbic system in the receptor preparations. Furthermore, a subcellular fractionation of tissue from the limbic system is used to preferentially extract the synaptosomes (41, 49). The limbic system described above (39) weighs, on the average, 15 grams in a dog brain. Following neurosurgical removal of the brain in the necropsy laboratory, the limbic system is neuroanatomically identified, excised rapidly in the cold, and the synaptosomal fraction isolated.

5.4.2 Preparation of Limbic Synaptosomes

The scheme in Figure 5.8 summarizes the subcellular fractionation of the limbic system tissue extract (49). After the brain tissue is homogenized in sucrose (0.32 M), centrifugation at 1000 x g and then at 17,000 x g produces a precipitate fraction which is resuspended and then separated in a discontinuous glucose gradient to yield three fractions. The middle layer, which disperses between 0.8 and 1.2 M sucrose, is diluted with water and centrifuged (100,000 xg) to produce the nerve ending particle fraction which is referred to as the synaptosomes. Synaptosomes are dissolved in Tris buffer (30 ml, 50mM) and then homogenized gently with a polytron. The preparation is left standing (1 hr; 0-4°C) and the synaptosome fraction swells. The synaptosomes are gently disrupted with a polytron. The synaptosomal fraction is analyzed for protein content and the original synaptosomal fraction (6.15 mg protein ml^{-1}) is divided into 30 equal portions (one ml each). One ml of the original fraction is added to one ml of Tris buffer (50 mM, pH. 7) and diluted to a 200 microliter volume to produce 0.615 mg protein as the original fraction to be used for radioreceptor assay.

5.4.3 Preincubation of receptors

The solution of synaptosomes containing a range of 0.5 to 2 mg protein ml^{-1} is utilized for RRA (32). The synaptosomal fraction is preincubated (1hr, 37°C) to destroy any remaining endogenous opioid peptide. The receptor preparation is incubated (15 min; 0°C) with a competing unlabeled drug or HPLC fraction of the expected enkephalin to effect binding of opioid peptide with the receptor. This incubation is followed by another incubation (2 hrs; 0°C) with tritiated dihydromorphine to competitively displace the peptide with this morphine analog. The incubation mixture is filtered (Whatman GF/B filters, 2.4 cm dia.) and then washed with Tris HCl (4 ml., cold), where recovery of radiolabeled compound optimized with five washings. The filter paper is dried with an infrared lamp, placed in toluene scintillation fluid (ten ml.), and radioactivity counted. It is important to wash out the

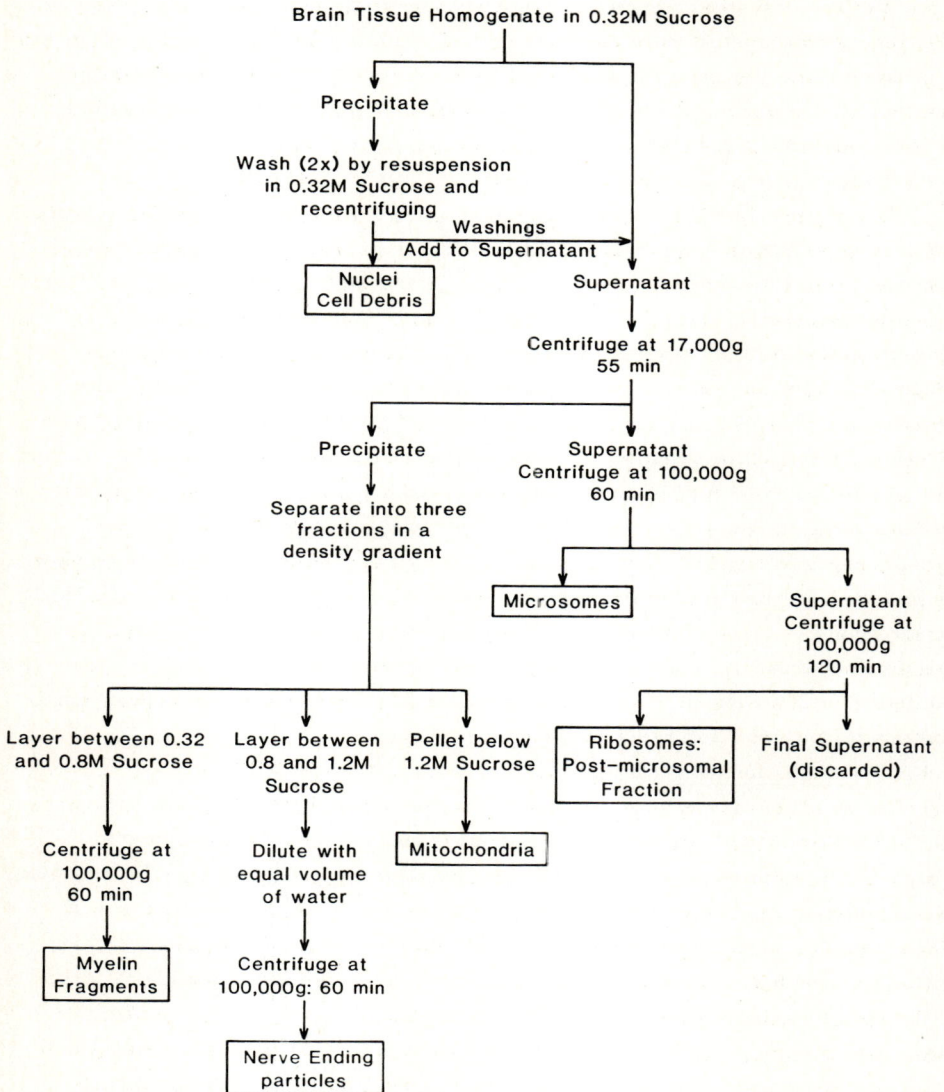

Fig. 5.8.   Subcellular fractionation of canine limbic system to isolate a
synaptosome preparation (49).

non-specifically distributed molecules before drying and counting of
radioactivity.

5.4.4 Development of receptor preparation

A preparation (4 mg protein) derived from whole limbic system
homogenate was tested for saturation (Figure 5.9) by increasing the amount
of tritiated dihydromorphine from zero to 5 mM and measuring the total,
specific, and non-specific binding. The Scatchard plot (B/F vs. B) of that
data is shown in Figure 5.10, where values of $K_a = 2.4 \times 10^9 \text{ M}^{-1}$ and $B_{max} = 15 \text{ fmol mg}^{-1}$ protein are calculated.

Two types of binding are normally observed in a receptor assay, specific
and non-specific binding. Specific binding is necessary for a receptor because
that receptor must exhibit binding to a unique ligand. Secondly, if the
receptor specifically binds to that ligand, then that receptor must be saturated
by increasing concentrations of that specific ligand to clearly demonstrate the
presence of a fixed number of receptors. Thirdly, the receptor must
demonstrate a high affinity to that specific ligand. These three criteria
define a specific radioreceptor to a selected endogenous ligand.

On the other hand, these tissue preparations also exhibit non-specific
binding which is experimentally defined by the phenomenon where other
ligands may be found which also bind to that selected receptor, but with much
lower affinity. That binding of a receptor to a non-specific ligand will not
be saturated and furthermore, the affinity of the receptor for that non-
specific ligand is much lower than that for the specific ligand. These
features are shown in Figure 5.9, where non-specific binding is shown as the
lower curve, total binding in the middle, and the difference between the two
binding levels, which represents specific binding, as the top curve (read on
right axis). Subtracting the non-specific from the total binding produces that
binding which is specific for tritiated dihydromorphine. The whole limbic
system preparation shown in Figure 5.9 does not display saturation
characteristics. A Scatchard plot (Fig 5.10) shows a linear decrease which
intercepts the abscissa at the maximum binding (B_{max}) at 15 fmol mg^{-1} protein
for this whole limbic system preparation.

To more efficiently achieve receptor saturation, the synaptosomal
subcellular fraction from the total limbic homogenate is selected to prepare
receptors (0.615 mg protein total). Again, this curve (Figure 5.11) is not
saturated, but it will be noted that, below 90 nmol of the tritiated
dihydromorphine, two straight lines occur in the Scatchard plot (Figure 5.12).
One straight line demonstrates a higher binding at 56 fmol mg^{-1} protein, and
the other line displays a lower binding at 160 fmol mg^{-1} protein. Most

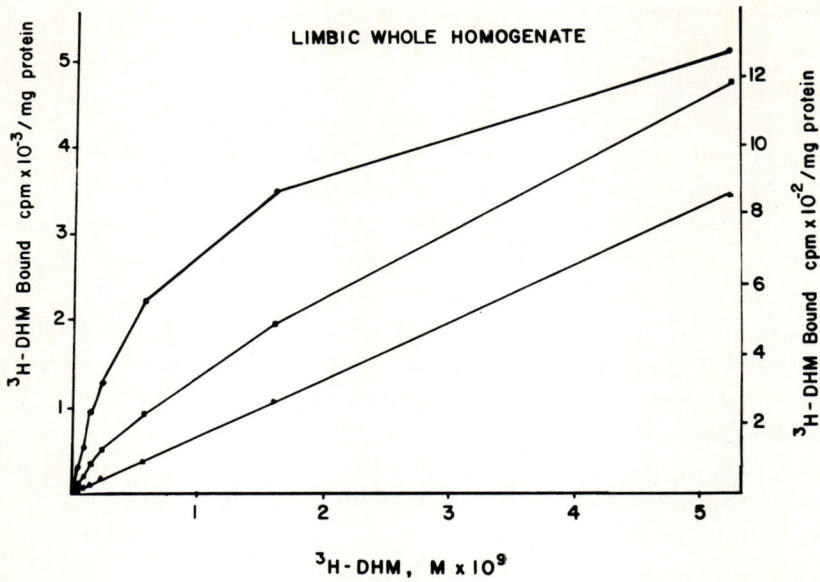

Fig.5.9. Receptor saturation analysis of whole limbic system homogenate (four mg protein total); (13).

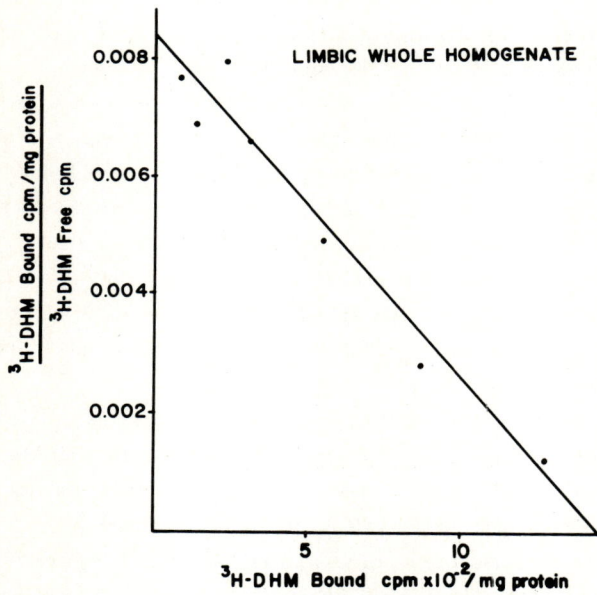

Fig.5.10. Scatchard plot of data in Figure 5.9 (13).

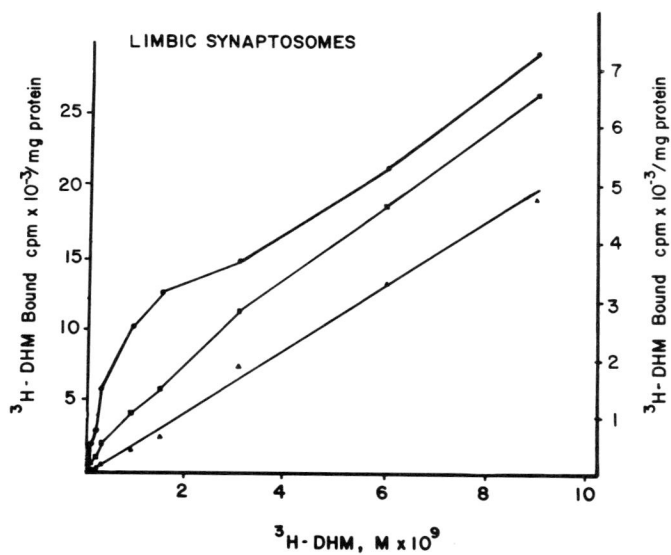

Fig. 5.11. Receptor saturation analysis of limbic synaptosomal fraction (0.615 mg protein); (13).

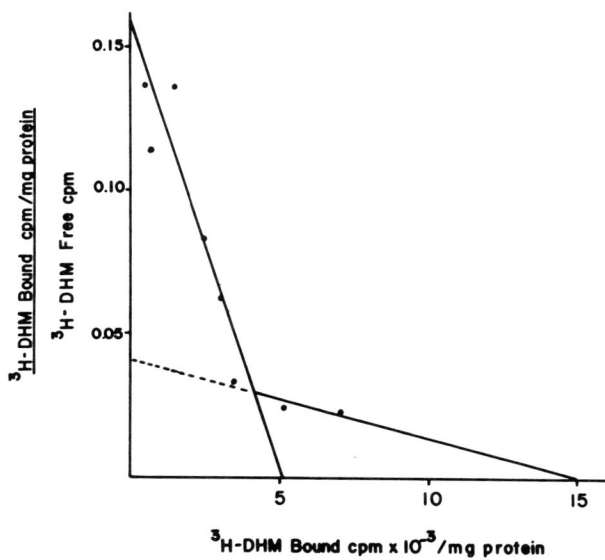

Fig. 5.12. Scatchard plot of data in Fig. 5.11 (13).

probably, these two B_{max} values represent the mu and delta receptors, respectively, for the canine limbic system synaptosomal preparation.

The non-specific binding level noted in the two previous saturation curves (Figures 5.9 and 5.11) was reduced in two ways. First, the number of filter washings was studied to most effectively remove any non-specifically bound ligand. It is found that five washings are optimal. Second, it was found that a preliminary wash of the glass filters with a solution of bovine serum albumin (0.5%, 30 sec.) minimizes non-specific binding of the peptides to the glass filter. These two parameters are included in the following experimental protocol.

When receptor saturation analysis is performed with the above experimental parameters on the same limbic synaptosomal preparation as above (Fig. 5.11), but with twice the amount of tritiated dihydromorphine, saturation is observed above 16 nM of tritiated dihydromorphine (Top curve, Figure 5.13). Furthermore, this preparation (0.123 mg protein total) demonstrates in the Scatchard plot (Figure 5.14), 29 and 70 fmol mg^{-1} of protein as B_{max} values for the high and low affinity receptors, respectively.

The straight lines in the Scatchard plots of Figures 5.10, 5.12, and 5.14 are tested for best-fit by a Hewlett Packard 97 calculator program. The statistical parameters for these three straight-lines are given in Table 5.4. Parameters for the best-fit of the straight lines given in the Scatchard plot headings are slope, intercept, and correlation coefficient (r^2).

TABLE 5.4. Statistical parameters for the straight lines shown in the Scatchard plots of Figures 5.10, 5.12, and 5.14.

Figure	Affinity Line	Slope	Intercept	Correlation coefficient (r^2)	No. Pts.
5.10	high	-5.8×10^{-6}	0.008	0.95	3
5.12	high	-3.1×10^{-6}	0.160	0.87	5
	low	-2.7×10^{-6}	0.004	0.79	3
5.14	high	-2.9×10^{-5}	0.078	0.92	3
	low	-5.6×10^{-6}	0.036	0.93	5

5.4.5 Radioreceptor Analysis

Figure 5.15 contains the RRA calibration curves (dotted lines) constructed by using a limbic system synaptosomal radioreceptor preparation where retention for tritiated dihydromorphine (% of control) is plotted versus molarity of added unlabeled ME and LE. Furthermore, a second limbic system synaptosomal radioreceptor preparation was made and the calibration curve (solid line) for ME is shown. Data in Figure 5.15 demonstrate that there is a

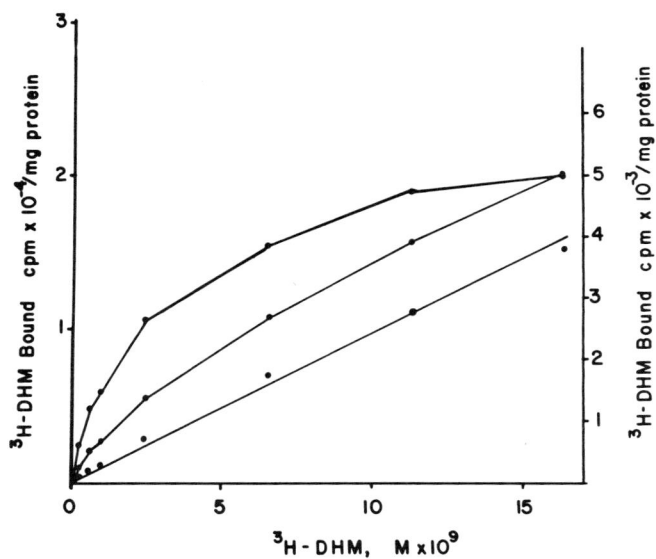

Fig.5.13. Receptor saturation analysis of limbic system synaptosomal fraction (0.123 mg protein total); (13).

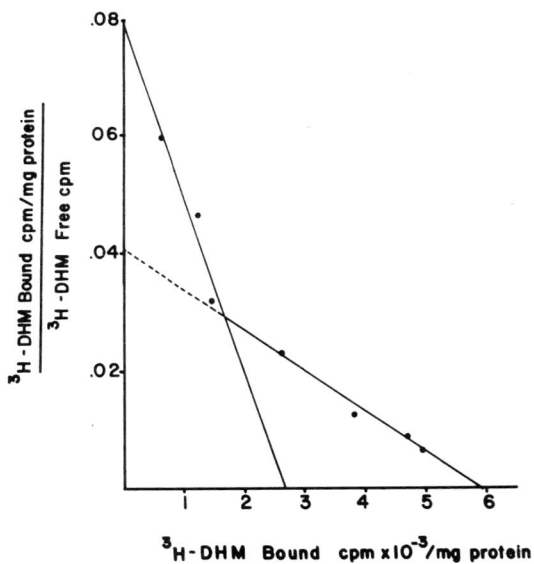

Fig.5.14. Scatchard plot of data in Fig.5.13 (13).

difference between the two canine limbic system synaptosomal preparations. For example, at the 100 nM level, a two-fold difference in concentration is shown by the 40% versus 20% displacement of the tritiated dihydromorphine from the opioid receptor. Nonetheless, for any one preparation and calibration curve, the data are sufficiently accurate.

Figure 5.16 contains the linearizing Hill plot, log M of unlabeled enkephalin vs. log [% inhibition/100-% inhibition], of the data shown in Figure 5.15.

5.4.6 <u>Measurement of Endogenous Enkephalins in Human Tooth Pulp Extracts</u>

One of the columns of data in Table 5.2 contains the analytical data calculated as ng of enkephalin RRA activity g^{-1} wet weight of human tooth pulp tissue. It must be remembered that this type of analytical data derives from the extent that tritiated dihydromorphine displaces the HPLC fraction from the receptor. No more structural information beyond that displacement phenomenon can be derived from this type of analytical measurement. The RP-HPLC fraction eluting at the retention time of synthetic ME for one patient's (DD) first tooth extracted was also studied by FAB MS to unambiguously confirm the presence of the opioid pentapeptide ME. An appropriate $[M+H]^+$ was visually observed on the MS oscilloscope at the nominal mass of 574, supporting the conclusion that ME is collected from the HPLC column at the appropriate retention time (5.2 mins.). Furthermore, material eluting at this retention time for patient DD tooth pulp was subjected to RRA, lending confidence to the fact that the target peptide is indeed the one being quantified by RRA.

The seven patients differ in the amount of receptor binding ME equivalents measured. Patient [TH(a)] has the lowest amounts, while another patient (JG) has the highest enkephalin-equivalent values for all four teeth. Patient JW has the highest value of enkephalin-equivalents for the first tooth extracted. RIA data are listed for comparison (12). It is clear from the data listed in Table 5.2 that an average of 3,668 nanograms of RRA enkephalin per gram of wet weight of tissue is found in the first tooth extracted from the four control patients (DD, JG, JW, and CB) whereas the three treated patients (TH, SG, and SS) have an average of 53 nanograms of RRA enkephalin per gram of wet weight tooth pulp tissue. These data represent a decrease of a factor by 69-fold between the control patients and the treated patients. This decrease in one selected opioid pentapeptide, ME, is not inconsistent with the working hypothesis developed for the endogenous opioid peptidergic systems in the tooth pulp. That working hypothesis states that a homeostatic relationship exists among the three opioid peptidergic families (endorphinergic, enkephalinergic, dynorphinergic), each of which is composed of larger

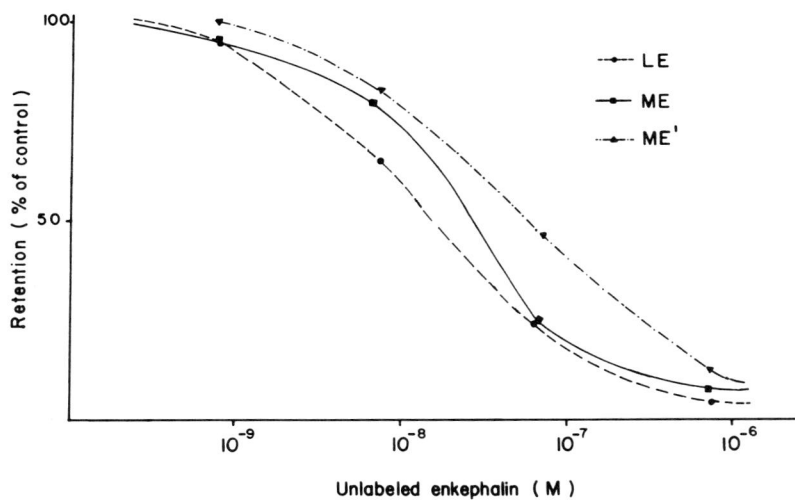

Fig.5.15. RRA calibration curve (13).

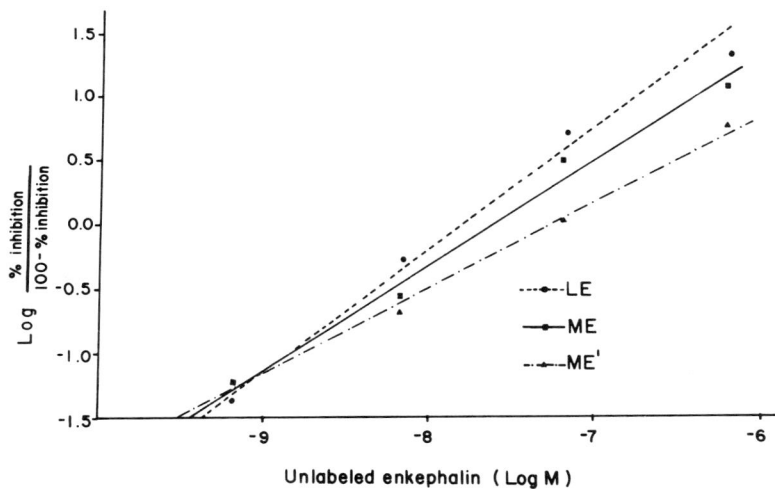

Fig.5.16. Hill plot of data in Fig.5.15 (13).

precursors leading to the "working" peptide which is rapidly metabolized. The second portion of the working hypothesis states that a noxious stimulus will activate a multi-step peptidergic process and then decreases of the relative concentrations of the enkephalins will occur. The enkephalins are hypothesized to interact with and decrease the firing rate of substance P-containing neurons (see Chapter 2). It must be remembered that we are focusing only on the opioid pentapeptides and not on all of the other possible peptides which may participate in the homeostatic relationship of the three peptidergic pathways noted above.

One of the experimental variables over which there is little control is the wet weight of the tooth pulp tissue derived from human teeth. However, it is noted that for the 26 teeth listed in Table 5.2, the average weight is .0674 grams ± 0.0315. The coefficient of variation (49%) is quite acceptable for the biologic source of this tissue.

In general, when a physiological experiment is undertaken, the amount of available tissue is apportioned into two fractions, with 10% used for RIA and 90% for RRA.

Clearly, RIA and RRA data do not either agree with each other nor with HPLC peak height data. While it might be expected that the three analytical modes should agree with each other, further reflection indicates that it is not necessarily the case. Conventional RP-HPLC measures only UV absorbance (or other parameter, depending on the detector) at one selected wavelength and cannot observe non-absorbing but yet, coeluting material which is usually present; RIA measures an Ab response to an oligopeptide, where that Ab is raised against a peptide-protein conjugate [in one case, 23 molecules of ME per ovalbumin molecule (5)]; and in RRA, a protein-lipid macromolecular complex on a synaptosomal surface is carefully isolated and used in a competitive binding assay versus tritiated dihydromorphine. It is crucial to realize that no rigid molecular proof is provided by, nor can ever be inferred from, any one of the three analytical methods HPLC, RRA, or RIA.

5.4.7 Conclusions

While a range of molecular specificities may be manifested in an RRA due to the presence of multiple opioid peptide receptors in this type of preparation, RRA data are used both as a convenient, rapid screen and, in conjunction with both RIA (which has a higher molecular specificity relative to a RRA) and the newly developed MS techniques (maximum molecular specificity), to measure endogenous opioids in biologic tissue and fluid extracts. RRA provides a faster (one day) analysis and an analysis which is amenable to a wider range of peptides that interacts with the opioid receptor.

A certain degree of variability is observed between canine synaptosomal receptor preparations. One preparation in Fig. 5.15 derives from one animal while the other preparation derives from pooling the limbic system from three animals. There may be various contributions to this variability – size, age, sex, species, and other unknown factors of the animal model. Nonetheless, the variability which is observed is no greater than a factor of two at the 100 nM level and decreases at both higher and lower levels. Such a small factor (two) is quite acceptable for this type of biological experiment.

RRA measurement of the two opioid pentapeptides ME and LE in human tooth pulp tissue extracts is presented and Scatchard plots suggest that two different (mu and delta) receptors may occur in these limbic synaptosomal receptor preparations.

Most significantly from an analytical viewpoint, the RRA provides another option in the total armamentarium available for analysis of endogenous opioid peptides. The available options include RP-HPLC, BA, RRA, RIA, and MS, where this list is arranged according to demonstrated and theoretically expected increasing molecular specificity of the analytical measurement process.

While MS can now be used to calibrate both the radioreceptor preparation and the RIA antibody (See chapters 6 and 7), once these two assays are calibrated, they can be used with increased confidence to provide a fast and facile assay method to monitor either whole tissue extracts or individual RP-HPLC fractions. RRA takes approximately one day to complete an analysis and this short time speeds up the number of physiological, electrophysio-logical, medical, dental, and drug studies that can be performed. These types of physiological studies are utilized to describe peptidergic (dynorphinergic, enkephalinergic, endorphinergic) pathways which are operative, participate in, or contribute to nociceptive mechanisms.

5.4.8   Radioreceptor assay screening

RRA is conveniently used for effective screening of chromatographically-separated opiates in human CSF (24-26, 31). For example, opiate peptides were purified from pooled human CSF (2 liters) and purification was effected by Sephadex G-10 and electrophoresis. HPLC and RIA were utilized for further characterization, and all procedures were guided elegantly by a specific RRA (31). Binding properties of opioid receptors and whole rat cerebrum were studied with tritium-labeled dihydromorphine, naltrexone, and LE as radioindicators (26). Three different kinds of opiate binding sites were observable.

Brain opiate receptors and endogenous opioid peptides were reviewed in terms of characterization and distribution (21). The highest concentrations of receptors were localized in the limbic system. As shown above, a synaptosomal preparation which is derived only from the limbic system has been used for RRA measurement of brain, CSF, and tooth pulp peptides (13).

High concentrations of opiate receptors were found in the anterior amygdala, the hypothalamus, and the medial thalamus (21). The latter location is important because of its participation in behavorial functions and processing of pain sensations. Moderate levels of receptors were found in the basal ganglia, mid-brain (especially periaqueductal gray) cerebral cortex, and the frontal, temporal, and hippocampal cortex. The basal ganglia location may be important for model studies in an effort to provide an understanding movement disorders such as Parkinsonism and Huntington chorea. Low levels were found in the lower brain stem and spinal cord, and no receptors were found in the cerebellum. It was interesting to note that no correlation with opioid receptors was found compared to the distribution of other known neurotransmitters (acetylcholine, serotonin, norepinephrine, dopamine, histamine, or GABA). This lack of correlation indicates probably that a separate "opiatergic" system exists. Two tritiated opiate ligands, diphrenorphine and etorphine, were especially useful for autoradiographic localization of opiate receptors in rat brain because of their high affinity for the opiate receptor and their low dissociation rates. These two properties minimize diffusion of these ligands away from the receptor site during histochemical processing of the tissue.

Even though the exact function of all the opiate receptors is not completely understood, certain pharmacological parameters of these receptors in relationship to opiate function cannot be overlooked. The limbic system correlates with many behavorial and emotional functions and, as noted above, a high level of opiate receptors is noted in the limbic system. These receptors may be involved in mediating the many behavorial effects of opiates (euphoria, mental cloudiness, memory changes, drowsiness) and the behavioral changes seen in morphine withdrawal states. It also has been noted that two kinds of enkephalins, ME and LE, may act on the mu and delta receptors. The association of the mu sites with ME and with the sensory systems supports the idea that peptides are included in opiate analgesia, whereas the delta receptors and LE are both associated with limbic structures and behavorial effects. In general, enkephalins are of a higher concentration in gray matter compared to white matter and are enriched in the synaptosomal fraction of a brain homogenate.

It has been noted that substance P coexists with both opiate receptors

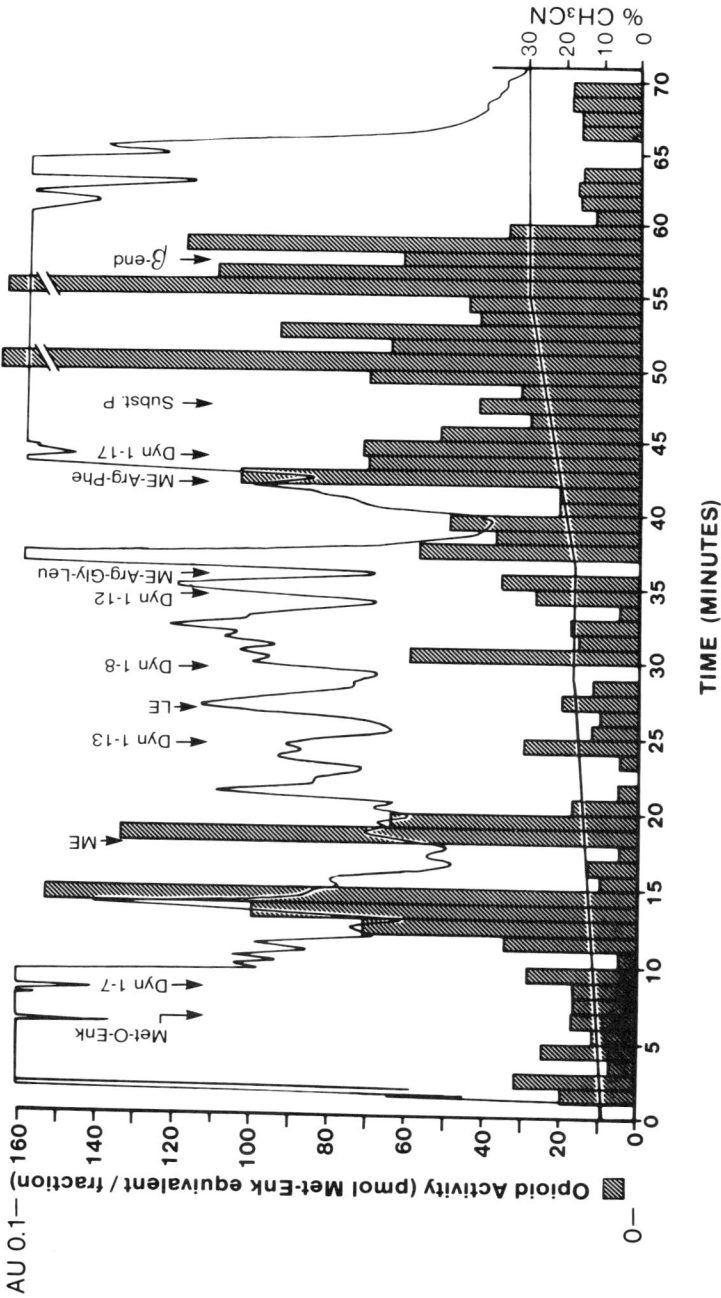

Fig. 5.17. Gradient RP-HPLC, RRA, and RIA of peptide-rich fraction from canine hypothalamus extract (50).

and enkephalins in the dorsal horn of the spinal cord. Substance P is thought to be an important neurotransmitter in primary afferent fibers subserving pain while opiate receptors are enriched in those fibers. Subsequently, it has been found that opiates and enkephalins decrease the firing rate of substance P neurons (Chapter 2) and block the release of substance P from their fiber terminals. Thus, a neuromodulatory mechanism which may be important for the analgesic action of opioids may be revealed.

5.4.9 <u>Screening for opioids via a combination of gradient RP-HPLC,</u>
       <u>radioreceptor analysis, and FAB MS</u>

A combination of gradient RP-HPLC, RRA, and FAB MS can be used effectively to assign an opioid activity profile to a biological tissue extract (50). A peptide-rich fraction from a tissue extract is passed through a C-18 Sep-Pak, is subjected to a gradient RP-HPLC chromatographic separation, and the individual fractions are then incubated with a limbic system synaptosomal receptor preparation, where constituents compete with the radioactive ligand for binding to the receptors. Figure 5.17 shows the opioid activity profile of the individual HPLC fractions measured as picomoles of ME equivalents. That RRA data is superimposed upon the corresponding RP-HPLC UV chromatogram of the peptide-rich fraction of a canine hypothalamic tissue extract. The gradient profile is shown, and the elution times for several synthetic peptide standards are indicated by the arrows along the top of the chromatogram. Three areas in this gradient chromatogram are the subject for analysis by FAB MS production of $(M+H)^+$ ions. In fraction 17, a fraction having a molecular weight, measured by FAB-MS, is noted at 1,259. Alpha-neoendorphin has a molecular weight of 1,229. By the use of MS peptide standards, we estimate that the peak at 1,259 mass units corresponds to 3-4 micrograms of material. This material will be further subjected to a linked-field (B/E) scan to elucidate amino acid sequence determining information. Fractions 21 and 22 show the presence of ME (600 nanograms), a large peak at 453, and a smaller peak at 485, where the latter two are presumed to be $(M+H)^+$ ions. This particular set of fractions indicates strongly the need to provide a separate detection system, even following HPLC separation where that detection system has an extraordinarily high level of molecular specificity. In other studies (8), we note that RRA analysis of this particular peak gives analytical data with highly variable results. Fractions 36 and 37 indicate the presence of an $(M+H)^+$ ion at mass 900, corresponding to a level of approximately 50 nanograms of the pro-enkephalin molecule ME-Arg-Phe. There is no dynorphin-6 observed at 1,571 mass units, but a very small and real peak at 1,755.

This type of combination analytical methodology is absolutely required when working with extracts of biologically complex materials such as brain peptides. While many workers presume that RP-HPLC provides a pure fraction, that result cannot be dependable because of the great number of peptides which are possible, coupled with the fact that conventional HPLC generally uses only one structural parameter of a molecule such as UV absorption, electrochemical activity, etc. Only by using a multiple set of detectors, for example, RRA, RIA, and FAB MS, can any level of confidence be attached to this type of separation methodology.

5.5 SUMMARY

This chapter reviews briefly three commonly used peptide assay methods- RIA, BA, and RRA. The areas of applicability, methods of preparation, basic principles, advantages, and disadvantages of each method are discussed. The focal point of this chapter is to increase the investigator's awareness of the potentially ambiguous molecular specificity which may be found when using a selected analytical method. Towards that end, the analytical methods are ranked in order of increasing molecular specificity: HPLC-UV, BA, RRA, RIA, and MS.

REFERENCES

1   S.A. Berson and R.S. Yalow in G. Pincus, T.V. Thimann and E.B. Astwood (Editors), The Hormones, Academic Press, N.Y., 1964, pp. 557-630.
2   S.A. Beron and R. Yalow, Adv. Biol. Med. Phys., 6 (1958) 349-430.
3   R.S. Yalow, Ann. Rev. Biophys. Bioeng., 9 (1980) 327-345.
4   D. S. Skelly, L.P. Brown and P.K. Besch, Clin. Chem., 19 (1973) 146-186.
5   C. Gros, P. Pradelles, C. Rouget, O. Bepoldin, F. Dray, M.C. Fournie-Zaluski, B.P. Roques, H. Pollard, C. Llorens-Cortes and J.C. Schwartz, J. Neurochem., 31 (1978) 29-39.
6   R.S. Yalow and J. Eng, Peptides 2 (1981) 17-23.
7   N. Brot, J. Werth, D. Koster and H. Weissbach, Anal. Biochem., 122 (1982) 291-294.
8   J.A. Walker, Jr., M.S. Thesis, Univ. Tenn., 1984.
9   F.S. Tanzer, D.M. Desiderio and S. Yamada, in D.H. Rich and E. Gross (Editors), Peptides: Synthesis, Structure, Function, Pierce Chem. Co., Rockford, Ill, 1981, 761-764.
10   M. Barber, R.S. Bordoli, G.J. Elliott and N.J. Horoch, Biochem. Biophys. Res. Comm., 110 (1983) 753-757.
11   P. Angwin and J.D. Barchas, J. Chromatogr., 231 (1982) 173-177.
12   F.S. Tanzer, D.M. Desiderio, C. Wakelyn and J. Walker, J. Dent. Res., submitted.
13   D.M. Desiderio, H. Onishi, H. Takeshita, F.S. Tanzer, C. Wakelyn, J. A. Walker and G. Fridland, J. Neurochem., submitted.

112

14 K. Diem and C. Lentner (Editors), Scientific Tables, Ciba-Geigy, Basle, 1970, p. 516.

15 D.W. Hoffman, R.A. Altschuler and J. Fex, Hear. Res., 9 (1983) 71-78.

16 D.M. Desiderio and M.D. Cunningham, J. Liq. Chromatogr., 4 (1981) 721-733.

17 R.L. Eskay and M.C. Beinfeld, Brain Res., 246 (1982) 315-318.

18 M.C. Al-Noaemi, J.A. Biggins, J.A. Edwardson, J.R. McDermott and A.I. Smith, Reg. Pept., 3 (1982) 351-359.

19 L. Olgart, B. Gazelius, E. Brodin and G. Nilsson, Acta. Physiol. Scand., 101 (1977) 510-512.

20 B.T. Hofreiter, J.P. Allen, A.C. Mizera, C.D. Powers and A.M. Masi, Steroids, 39 (1982) 547-555.

21 S.F. Atweh, in Smith and Lane (Editors), Neurobiology of Opiate Reward Processes, Elsevier Biomedical Press, Amsterdam, 1983, pp. 59-87.

22 R. Simantov and S.H. Snyder, in Opiates and Endogenous Opioid Peptides, H. Kosterlitz (Editor) Elsevier (1976) 41-48.

23 L. Terenius and A. Wahlstrom, Acta Physiol. Scand., 94 (1974) 74-81.

24 L. Terenius and A. Wahlstrom, Life Sci., 16 (1975) 1759-1764.

25 A. Wahlstrom, L. Johansson and L. Terenius, in Opiates and Endogenous Opioid Peptides, H. Kosterlitz (Editor), Elsevier, Amsterdam, 1976, 49-55.

26 L. Terenius, Psychoneuroendocrinology, 2 (1977) 53-58.

27 L.V. Knorring, B.G.L. Almay, F. Johansson and L. Terenius, Pain, 5 (1978) 359-365.

28 F. Nyberg and L. Terenius, Life Sci., 31 (1982) 1737-1740.

29 J.F. McGinty, S.J. Henriksen, A. Goldstein, L. Terenius, and F.E. Bloom, Life Sci., 31 (1982) 1797-1800.

30 S. Borg, H. Kvande, U. Rydberg, L. Terenius and A. Wahlstrom, Psychopharmacology, 78 (1982) 101-103.

31 F. Nyberg, A. Wahlstrom, B. Sjolund and L. Terenius, Brain Res., 259 (1983) 267-274.

32 E.J. Simon, J.M. Hiller and I. Edelman, Proc. Nat. Acad. Sci., 70 (1973) 1947-1949.

33 J.A.H. Lord, A.A. Waterfield, J. Hughes, and H.W. Kosterlitz, in Opiates and Endogenous Opioid Peptides, H.W. Kosterlitz (Editor), Elsevier, Amsterdam, 1976, 275-280.

34 M.E. Lewis, M. Mishkin, E. Bragin, R. M. Brown, C.B. Pert and A. Pert, Science, 211 (1981) 1166-1169.

35 K. Stengaard-Pedersen, Acta Pharmacol. Toxicol., 50 (1982) 213-220.

36 V. Pliska, J. Receptor Res., 3 (1983) 227-238.

37 E.A. Barnard and C. Demoliou-Mason, Brit. Med. Bull., 39 (1983) 37-45.

38 S.F. Atweh and M.J. Kuhar, Brit. Med. Bull., 39 (1983) 47-52.

39 L. Heimer, The Human Brain and Spinal Cord, Springer-Verlag, New York, 1983, pp. 327-329.

40 J.A.H. Lord, A.A. Waterfield, J. Hughes, and H.W. Kosterlitz, Nature, 267 (1977) 495-499.

41 K.J. Chang and P. Cautrecasas, J. Biol. Chem., 254 (1979) 2610-2618.

42 K.J. Chang, B.R. Cooper, E. Hazum and P. Cuatrecasas, Molec. Pharm., 16 (1979) 91-104.

43 A.P. Smith and H.H. Loh, Horm. Prot. Peptides, 5 (1981) 89-159.

44 J.L. Vaught, R.B. Rothman, and T.C. Westfall, Life Sci., 30 (1982) 1443-1455.

45 B. Tabakoff and P.L. Hoffman, Life Sci., 32 (1983) 197-204.

46 R. Schulz, M. Wuster, and A. Herz, Peptides, 3 (1982) 973-976.

47 S.J. Paterson, L.E. Robson, and H.W. Kosterlitz, Brit. Med. Bull., 39 (1983) 31-36.

48 R. Dubner, B.J. Sessle and A.T. Storey, Neural Basis of Oral and Facial Function, Plenum, N.Y. 1978, 483 pp.

49   E.G. Gray and V.P. Whittaker, J. Anat., 96 (1962) 79-88.
50   D.M. Desiderio, H. Takeshita, C. Dass, and G. Fridland, in preparation.

Chapter 6

MASS SPECTROMETRY OF PEPTIDES

6.1 INTRODUCTION

Today, the most efficient and effective MS method to ionize a peptide
is FAB MS. Furthermore, amino acid sequence-determining information may be
obtained by using unimolecular decompositions, CAD, MIKES, and linked-field
methods. This chapter will briefly review those salient features of MS
which are pertinent to peptide research.

In principle, analytical methodology includes both qualitative analysis
and quantitative analysis. On one hand, the structure of an unknown
compound may be elucidated using qualitative analytical techniques, where
X-ray crystallography and MS are two instrumental methods that yield the
most molecular structure information for a pure compound. MS offers special
advantages for structural elucidation by providing data obtained in the high
resolution – accurate mass – elemental composition mode from nanograms of
pure compound. On the other hand, for quantitative analysis, RIA and MS
are considered to be the two analytical methods which offer the highest
level of detection sensitivity. One of the themes of this book is to
demonstrate the fact that optimal molecular specificity is obtained only in
the case of MS. Because MS plays such a significant role in both qualitative
and quantitative analyses of biologically important compounds, and especially
peptides, the objective of this chapter is to review those relevant features
of MS which apply to peptide analysis.

The number of significant developments in MS has grown over the past
decade as perhaps for no other analytical technique. This chapter will not
attempt to review the entire field or history of MS in either the breadth or
depth of available material, but rather will introduce to a newcomer the
basic principles of those aspects of MS, including EI, CI, FD, FAB, CAD, and
linked-field scanning MS that especially pertain to peptide analysis.
Applications of these mass spectrometric techniques to analysis of other
classes of compounds are readily derived from the literature which is listed
in this chapter. Plasma desorption, direct chemical ionization (DCI), triple
stage quadrupole (TSQ) MS, and Fourier transform MS (FTMS) are other areas
that will be mentioned only briefly.

A wide range of reference books is available to the newcomer to the field of MS and will serve to review the various types of MS techniques that are mentioned in this chapter. The main purpose of this chapter is to distill from the current literature those selected aspects of MS which impinge directly upon the unique combination of the three fields which are developed in this book - neuropeptide research, RP-HPLC, and MS analysis of peptides. In other words, this book represents a highly focused treatise of these three selected fields.

Several books are available for basic interpretation techniques in MS (1-7). Several MS conferences are held on a regular basis and include the annual conference of the American Society of Mass Spectrometry (8), the Triennial International Mass Spectrometry Conference (9-11), and other national conferences (Britain, Japan, Belgium, and Australia). A quarterly review of MS applications (12) and a biennial review of MS (13) are published.

MS methodology has been applied to a variety of research topics including biochemistry, medicine, neurobiology, organic acidurias, amongst others (14-18).

Two books devoted to biochemical applications of MS have been published (19,20). Other selected books collecting various MS techniques and edited by experts in those respective fields include basic instrumentation and theoretical treatises (21), metastable ions (22), collision spectroscopy (23), chemical applications of high performance MS (24), topics in organic MS (25), soft ionization biological MS (26) and tandem MS (27). Several books and reviews are available on the use of stable isotopes (28,29) and experimental aspects of quantitation (30). Several other helpful books of an ancillary nature to MS are available and include a very useful compilation of published chemical derivatization reactions (31), a useful list of impurities encountered in MS (32), a table of molecular weights to be used with MS (33), and a table of accurate masses and isotopic abundances for use in MS (34). Other volumes available to the researcher include basic FI MS (35) and computers in MS (36). This collection of reference books available to those interested in the above-listed aspects of MS is not an exhaustive list. There are other, more specialized volumes also available in a selected field to assist a researcher in developing their specific applications of MS.

The purpose of this chapter is to facilitate the learning experience of a newcomer to the field of MS (see Fig. 1.1). Towards that end, this chapter will include a brief review of EI and CI techniques followed by an in-depth discussion of the two newer ionization methods of FD and FAB MS methods, where the latter two are considered alone and coupled with the techniques of CAD and linked-field scan analysis.

6.2 ELECTRON IONIZATION MASS SPECTROMETRY OF PEPTIDES

Over 70 years ago, two isotopes of neon were separated with a positive ray parabolic MS apparatus. For many years following that initial experiment, MS instruments were used mainly by physicists in research laboratories until oil chemists started to use EI MS techniques for matrix analysis of volatile petroleum products, where low resolution MS instruments with magnetic field deflection were used to produce digitized data output. When researchers in chemistry laboratories began utilizing MS around the 1960's for structural elucidation of organic compounds, it was quickly realized that chemically derivatized peptides are amenable to MS (37, 38).

In EI, a heated rhenium filament is used to emit several hundred microamperes of electron current which is collected over a voltage gradient of generally 70 electron volts. Ions are formed by electron impact with gaseous molecules, where either an electron is removed from the molecule to produce a positive ion-radical species or a negative ion-radical is generated. These ions undergo fragmentation involving cleavage and/or rearrangement processes which occur in the ion source within picoseconds to microseconds following ionization. The ions are extracted from the ion source of a sector instrument by a few keV accelerating voltage. The ions undergo direction-focusing mass analysis with a magnetic field in a single-focusing instrument while, in a double-focusing instrument, an electric sector additionally performs velocity-focusing. In a double-focusing instrument, mass can be determined with an accuracy to within a few parts per million (ppm), and that accurate mass can be converted with appropriate computer programs to corresponding elemental composition data (39-42).

Figure 6.1 contains a scheme of the two types of double-focusing mass spectrometers. In the Nier-Johnson geometry, the ion beam is focused to a focal point, while in the Mattauch-Herzog geometry, to a focal plane. Magnetic field scanning can be done with both MS geometries to produce a mass spectrum, while in the latter case, a photographic plate can be used instead to record simultaneously the entire mass spectrum (40-42).

6.2.1 Chemical Derivatization of Peptides

At first, the polar peptide molecule is not considered to be readily amenable to MS analysis because, before either EI or CI can occur, the molecule must first be vaporized. Several structural features of a peptide molecule contribute to its polarity - the carbonyl and amide groups in the same or two different peptide chains can readily hydrogen-bond to each other;

there are zwitterionic charges in a peptide chain due to the presence of both the protonated N-terminus cation and the C-terminus carboxylate anion; and polar side-chains (K, R, D, E) may be present. Chemical derivatization techniques have been developed and include reduction of the amide bond (1, 38, 43-45) and the combination of acylation-esterification-permethylation (46-49) reactions. The chemically derivatized peptides have sufficient thermal stability conferred upon them so that they can be vaporized efficiently without chemical degradation; dipeptide (50) and oligopeptide (51) derivatives are also amenable to GC separation.

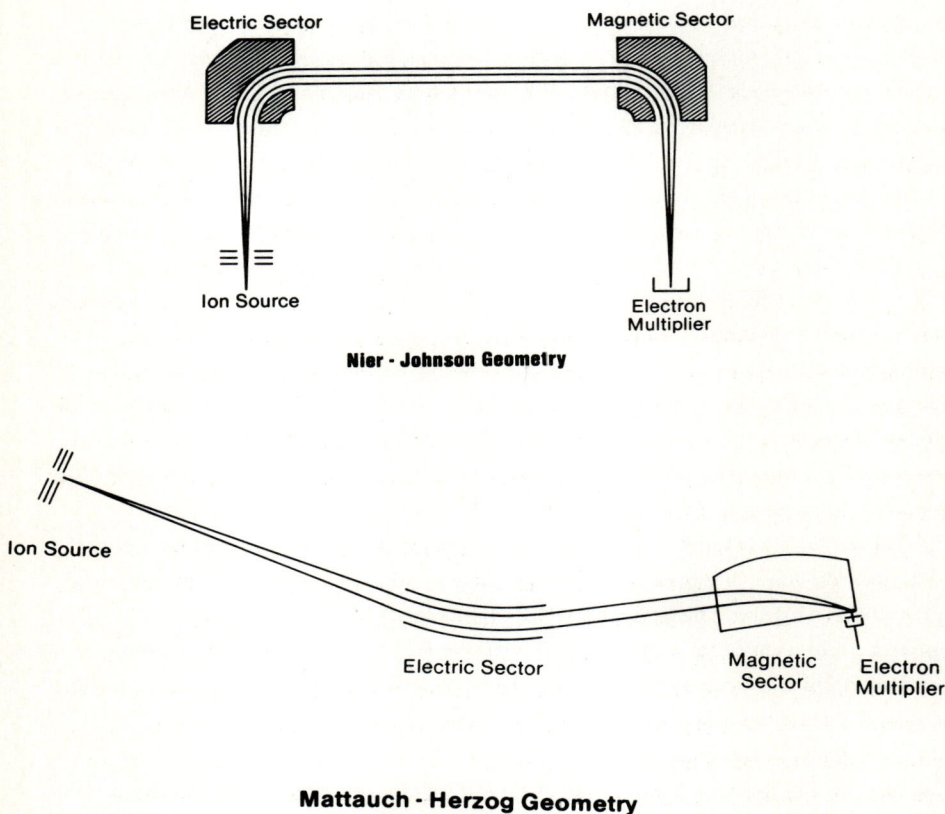

Fig. 6.1. Nier-Johnson and Mattauch-Herzog double-focusing mass spectrometer geometries.

6.2.2 Types of Mass Spectra of Peptides

Generally, the abundance of the molecular ion-radical ($M^{+\cdot}$) of a derivatized peptide in an EI mass spectrum is quite small in terms of relative intensity, if it is present at all. The method by which the amino acid sequence of an unknown peptide is determined from its corresponding mass spectrum is more readily understood by considering the following general fragmentation scheme, where amino acid sequence-determining ions are produced by fragmentation of the peptide amide bond. Other types of ions are also produced. For example, for a peptide containing n amino acids, the n amino acid sequence-determining ions, which includes $M^{+\cdot}$, can be used to define the amino acid sequence of the peptide. The subscript n in the following scheme denotes the nth amino acid, counting from the appropriate terminus which is indicated by N or C.

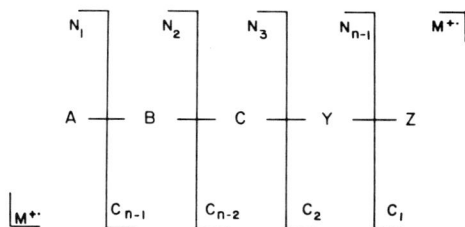

$$
\begin{array}{ccccccccc}
N_1 & & N_2 & & N_3 & & N_{n-1} & & M^{+\cdot} \\
\hline
A & + & B & + & C & + & Y & + & Z \\
\hline
M^{+\cdot} & & C_{n-1} & & C_{n-2} & & C_2 & & C_1
\end{array}
$$

In general, other fragmentation processes also occur and, in some fortunate cases, amino acid sequence-determining ions overlap from the two directions. In order to exemplify the type of MS data obtained from a peptide, Figure 6.2 contains the low resolution EI mass spectrum of the permethylated tetrapeptide 2-pyrrolidone-5-carboxyl-prolyl-tyrosyl-histidine -amide, abbreviated as PCA-Pro-Tyr-His-NH_2 (52). The EI-induced fragmentation pattern (Fig. 6.3) of this permethylated tetrapeptide demonstrates amino terminal-containing ions at masses 98, 126, 195, 223, 386, 414, 551, 559, and $M^{+\cdot}$ at 623. Carboxy terminal-containing ions occur at masses 44, 72, 209, 237, 400, 428, 497, 525, and again, 623. This mass spectrum is one of the most complete examples demonstrating the two directions of overlapping amino acid sequence-determining ions. Other ions included in the mass spectrum include the losses of the side chains, where these ions are found at masses 95 (His) and 121 (Tyr), as well as other fragmentation and rearrangement ions.

Computer programs have been written for the fast, facile, and objective determination of the amino acid sequence of a peptide which is derived from

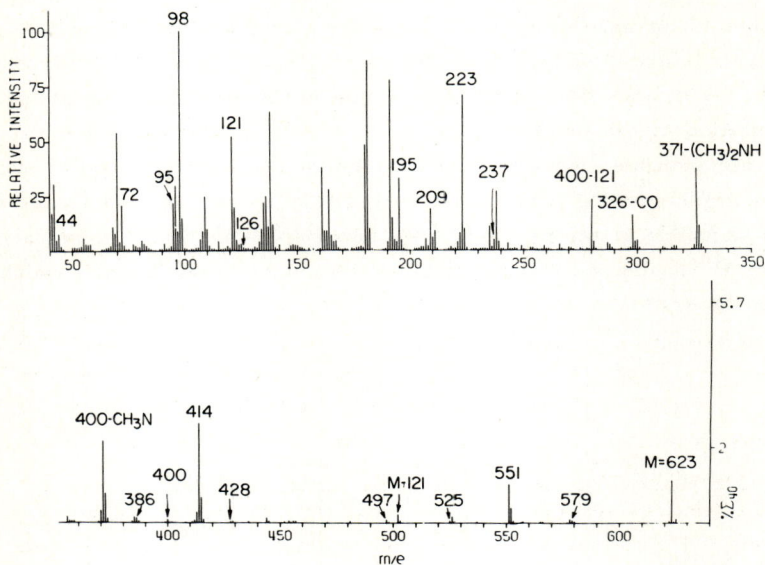

Fig. 6.2. Mass spectrum of permethylated PCA-Pro-Tyr-His-NH$_2$ (52).

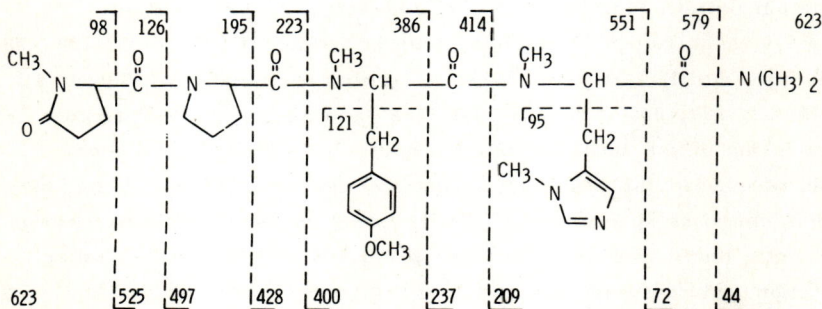

Fig. 6.3. Fragmentation pattern of PCA-Pro-Tyr-His NH$_2$ (52).

high resolution EI MS data (53-54). Several reviews are available on the MS of peptides (55-56).

The structure of an extremely potent biologically active brain peptide was elucidated with the aid of MS (57-58). The determination of the amino acid sequence of the hypothalamic thyrotropin releasing factor (TRH) tripeptide (See Chapter 2) is one of the first cases wherein the two peptide termini are blocked and the presence of proline inhibited carboxypeptidase digestion (Fig. 6.4). Blocked termini continue to be a prevalent structural feature of several brain peptides (See Table 2.1). The amino acid sequence of the TRH peptide is pyroglutamyl-histidyl-prolineamide. At the time of analysis, no proteolytic enzymes were available to hydrolyze this peptide. Since then, a PCA-ase was found. High resolution MS provided accurate mass and elemental composition data for ready interpretation of the amino acid sequence information (Fig. 6.5). The EI mass spectra were corroborated with CI MS data (58). This amino acid sequence determination was one of the first cases where a peptide of limited volatility was introduced directly into the electron beam of a mass spectrometer to facilitate ionization.

6.2.3   Advantages of Electron Ionization Mass Spectrometry

EI MS offers several significant experimental advantages to the peptide chemist. For example, in the most favorable cases, it is possible to completely determine, within minutes, the amino acid sequence of an unknown peptide in one mass spectrum from nanogram amounts of a chemically derivatized peptide. Furthermore, if the N-terminus and/or C-terminus is chemically blocked (which is the case with many endogenous brain peptides), MS offers a significant advantage over other methods because one chemical derivatization step is bypassed.

On the other hand, disadvantages are always present for MS sequencing of peptides in the EI and CI modes. For example, chemical derivatization is required. Polar side chains such as K and R may present some experimental difficulties. Some of the derivatization chemistry is relatively complex, such as proper preparation of the dimethylsulfinyl carbanion catalyst utilized for permethylation (59). Generally, little ion current is found for the molecular ion. An instrumental disadvantage is that, until recently, most commercially available MS instruments had a molecular weight limit of ca. 1000 mass units, limiting analysis to a nonapeptide. However, FAB and laminated high field magnets are now available to overcome these limitations.

MS information is a complementary technique which can be used in conjunction with automated peptide sequencers. In some cases, solubility problems may be experienced in a peptide sequencer and MS offers an important adjunct, or necessary alternative method.

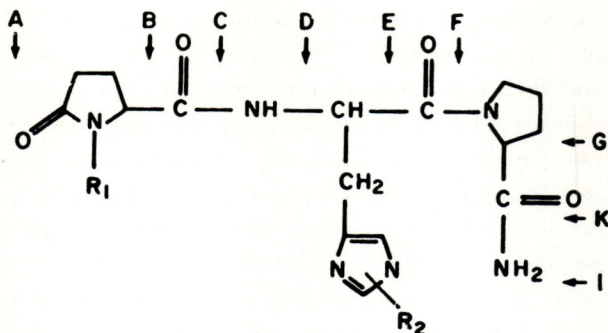

Fig. 6.4. Fragmentation points derived from the mass spectrum of TRF (57).

Fragmentation Points	Measured Mass	Error mMU	Elemental Comp.
A → B	84.0463	+1.5	C_4H_6NO
A → B, R_2 = CH_3	98.0576	−3.0	C_5H_8NO
D → E	94.0535	+0.4	$C_5H_6N_2$
D → E, R_2 = CH_3	108.0673	−1.4	$C_6H_8N_2$
F → K, minus 4H	93.0211	−0.4	C_5H_3NO
F → I, minus 3H	110.0480	0.0	$C_5H_6N_2O$
G → I	44.0120	−1.6	CH_2NO
A → E, R_1 or R_2 = CH_3, minus H or C → K or D → I, minus H	234.1112	−0.4	$C_{11}H_{14}N_4O_2$
C → I, minus 2H	248.1147	0.0	$C_{11}H_{14}N_5O_2$
A → E, R_1 = R_2 = CH_3 or D → I, R_2 = CH_3 or C → K, R_2 = CH_3, minus H	249.1327	−2.4	$C_{12}H_{17}N_4O_2$

Fig. 6.5. Accurate mass and elemental composition data for TRF (57).

6.3   EXAMPLES OF ELECTRON IONIZATION MASS SPECTRA OF
      CHEMICALLY DERIVATIZED PEPTIDES

6.3.1 Reduced peptides

   Lithium aluminum hydride is utilized for chemical reduction of the
peptide amide bond to produce polyamino alcohols in a procedure to confirm
complementary DNA sequencing information (60). Basically, random acid
hydrolysis digests the long peptide into shorter peptide fragments, the
peptide amide bond is reduced to a polyaminoalcohol with lithium aluminum
hydride (or deuteride), the polyaminoalcohol is trimethylsilylated, and
GC-MS is used to chromatographically separate the mixture of peptide
hydrolysis fragments into its constituents and to obtain the mass spectrum
and then structure of each one of the derivatized polyaminoalcohols. Peptide
fragments containing up to six amino acid residues are amenable to this
methodology.

   Borane has been used to provide a gentle procedure to effectively
reduce the amide bond in peptides of interest (44). The mass spectra are
clean, and structurally informative ions are present. Eleven peptides
(dipeptides and dipeptide amides) were studied where trifluoroacetylated,
pentafluoropropionylated, and heptafluorobutyrylated derivatives were used.
In all cases, structurally significant ions were noted.

6.3.2 Permethylated peptides

   One of the first examples of the use of MS of peptides was a study of a
naturally-occurring acylated and methylated peptide extracted from
Mycobacterium fortuitum (47). M. fortuitum produces a peptidolipid
(fortuitum) which is comprised of a mixture of two acylating groups (20
and 22 carbons) on a nonapeptide which contains several N-methyl amino
acid residues:

$CH_3-(CH_2)_{18,20}CO$-Val-MeLeu-Val-Val-MeLeu-(Ac)Thr-(Ac)Thr-(Ac)Thr-Ala-Pro.

   The two molecular ions were found at masses 1331 and 1359,
respectively. This work prompted a series of chemical studies to devise
a silver oxide catalyst to readily effect permethylation of oligopeptides
(46). It was then found that this heterogeneous reaction mixture could be
effectively replaced with a homogeneous catalyst, the dimethylsulfinyl
carbanion (61), which is conveniently produced by reacting sodium hydride
with dimethyl sulfoxide (59). This carbanion efficiently removes all
hydrogen atoms which are not attached to a carbon atom in the peptide
molecule. The highly reactive peptide polyanion is treated with an
alkylating reagent such as methyl iodide and the resulting hydrophobic

permethylated peptide is extracted into chloroform. While longer chain
alkyl iodides also work well as derivatives, it was found that a number of
undesirable rearrangements occur in their mass spectra to complicate their
analytical utility.

Trideuterated-methyl iodide is a useful reagent to shift those ions
containing a methyl group and also to readily facilitate the rationalization
of ion geneses (48).

This type of detailed structural analysis used to readily rationalize the
genesis of significant peaks occurring in the mass spectrum of a peptide is an
important prelude to two types of analytical studies. On one hand, it is
important to understand in sufficient structural detail the fragmentation modes
which are available to a peptide to be able to elucidate with MS methods the
structure of an unknown peptide. On the other hand, unequivocally knowing
the structure of a fragment ion is a crucial requirement demanded for
measurement of a peptide in a biological extract (See Chapter 7). Towards
these two ends, ions labelled n (mass 213), o (199), p (339), and q (509)
occurring in the mass spectrum (Fig. 6.6) of a derivatized octapeptide, N,
O-permethyl, N-acetyl KSAQQGGY-NH$_2$ were studied in detail with stable
isotope-labeled chemical derivatives and their corresponding structures are
shown here as four selected examples of this needed structural detail.

The permethylation chemical reaction is quite suitable for use in MS
because these derivatives fulfill several vaporization requirements: decrease

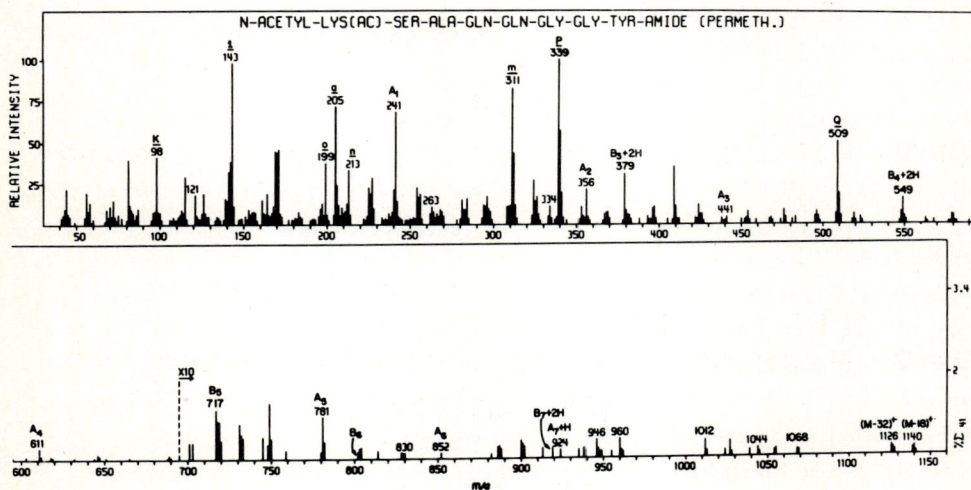

Fig. 6.6.   EI mass spectrum of N-peracetyle, N, O - permethylated
KSAQQGGY-NH$_2$ (48).

a, m/z 213

(Shifts to 216 with d₃ – acetylation; to 222 with d₃ – methylation; and 225 with d₃ – acetylation plus d₃ – methylation).

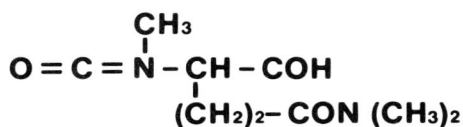

o, m/z 199

(Shifts to 200 with d₃ – methylation; no shift with d₃ – acetylation).

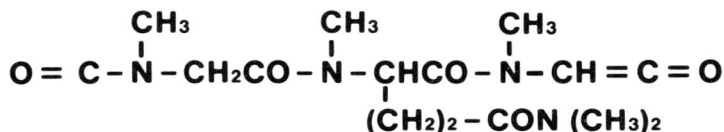

c, m/z 339

(Shifts to 354 with d₃ – methylation; no shift with d₃ – acetylation).

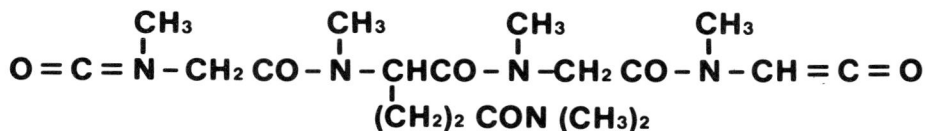

d, m/z 509

(Shifts to 527 with d₃ – methylation; no shift with d₃ – acetylation).

the zwitterionic character of the peptide, increase the volatility which is
lowered by interchain hydrogen bonding; form appropriate chemical
derivatives of those amino acids that contain functional groups- a process
which will also conveniently remove a charge; and, in some cases, stabilize
particular fragmentation pathways which optimize amino acid sequence-
determining information. The permethyl derivative is well-suited for MS
purposes because only 14 mass units per replaceable hydrogen are added to the
molecular weight of the original peptide. This minimal increase in mass is an
important consideration whenever a large molecular weight and/or a large
number of replaceable hydrogens is involved in a peptide molecule so that the
usable mass range of the mass spectrometer is not exceeded.

A general scheme for the efficient permethylation of a peptide was
developed, where the molar quantities of methyl iodide and
methylsulfinylmethide carbanion were both in a tenfold excess of the number
of peptide equivalents which, in turn, is equal to the number of moles of
peptide times the number of chemically replaceable hydrogens. For example,
to methylate 50 micrograms (A nanomoles) of a peptide with a molecular weight
of M and containing B available methylation sites, it is necessary to add
10 x A x B nanomoles of carbanion and an equivalent number of nanomoles of
methyl iodide. In practice, for example, 15 micrograms of any peptide may be
conveniently permethylated by adding eight microliters of a carbanion solution
of approximately one M concentration, plus 500 nanoliters of methyl iodide.

A short reaction time for permethylation was devised subsequently (62)
for amino acid sequence determination.

6.3.3 Enzymology/gas chromatography-mass spectrometric studies of peptides

Effective use is made of GC purification of a mixture of peptides which
has been chemically reduced and trimethylsilylated. Investigators have taken
advantage of a series of proteases known as dipeptidylaminopeptidases (DAP)
to cleave peptides to smaller fragments (63,64). These peptidases (cathepsins)
are found in brain. A peptide is treated with DAP I to attack the N-terminus
of the peptide chain and consecutively cleave sequential dipeptides until
either the protease-inhibiting amino acid residue proline or the C-terminus
is encountered. DAP I does not hydrolyze beyond a proline residue, whereas
DAP IV does. The mixture of dipeptides produced by DAP proteolysis is
chemically derivatized and subjected to GC-MS analysis. After the amino
acid sequence of the constituent dipeptides is determined in the mixture,
another fraction of the original intact peptide is subjected to one cycle
of Edman degradation to remove only the first amino acid residue. The

resulting des-amino terminal peptide is also subjected to DAP treatment to provide a second series of dipeptide sequences from which the structure of the original peptide is deduced by overlapping the two obtained peptide structures.

Advantage was taken of DAP I to develop a simple straightforward analytical method for differentiating Asp/Glu-containing dipeptides, respectively (64). The technique involves two separate chemical methods to form methyl esters. The first method employs 0.1 N HCl/MeOH at room temperature for four hours to reveal all dipeptides except the non-volatile Asn- and Gln-containing dipeptides. The second procedure (0.1 N HCl/MeOH, 45°, 16 hours) solvolyses the Asn/Gln residues, and new Asp- or Glu-containing dipeptides occur.

6.4 CHEMICAL IONIZATION MASS SPECTROMETRY

The basic principles of CI MS are reviewed elsewhere (65). A relatively high pressure (0.1 - 1.0 torr) is produced in the ion source of the mass spectrometer with a reagant gas (methane, isobutane, ammonia, etc.). Ionization of methane induces a series of ion-molecule reactions whereby a set of reagant ions (CH_5^+, $C_2H_5^+$, and $C_3H_5^+$) results. In the gas phase, this set of protonated hydrocarbon molecules acts as strong Brønsted acids to effectively ionize a compound of interest, according to that compound's gas phase basicity. The CI process is less energetic compared to EI. Because less energy is transferred to the compound, less fragmentation generally results. Furthermore, fragmentation can be controlled by the exothermicity of the CI reaction. In general, abundant $(M+H)^+$ ions are formed in CI and are used for quantification of many molecules.

Peptides (66) and amino acids (67) have been studied by CI MS. In the latter case, free amino acids and their amides, methyl esters, and N-acetyl derivatives were studied.

6.5 FIELD DESORPTION-MASS SPECTROMETRIC STUDIES OF PEPTIDES

Noticeable advancements were made in peptide analysis following the development of FD MS (68a). For the first time, relatively large peptides having a molecular weight up to 1,000-1,500 mass units are amenable to MS analysis because those peptides are ionized by FD. Experimentally, it is very significant to realize that no derivatization chemistry is required for FD MS, and this fact is quite important for studies of endogenous peptides. The FD mass spectra of peptides are quite simple when compared to both EI and CI mass spectra. To illustrate this spectral simplicity and to

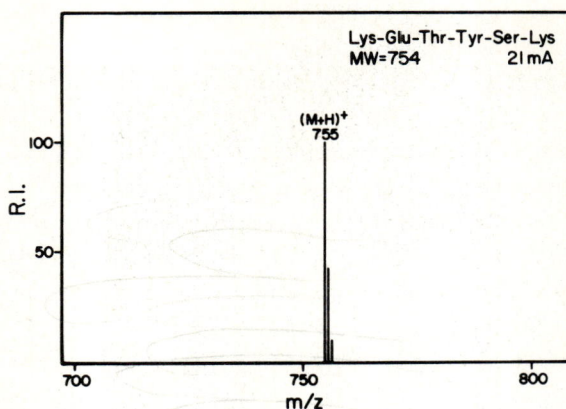

Fig. 6.7. FD mass spectrum of the hexapeptide KETYSK (70).

point to the use of FDMS as an analytical method to measure peptides, Figure 6.7 contains the low resolution FD mass spectrum of synthetic model hexapeptide Lys-Glu-Thr-Tyr-Ser-Lys (KETYSK) with a molecular weight of 754.

The FD mass spectrum of this polar hexapeptide molecule contains only the $(M + H)^+$ ion at 755 and associated isotope peaks. In general, little or no amino acid sequence-determining information is produced by FD ionization processes. On one hand, production of large ion currents corresponding to only the $(M + H)^+$ ion of a peptide readily provides the analytical basis to measure endogenous peptides in biologic tissue extracts. This significant point will be discussed in greater detail in Chapter 7. On the other hand, amino acid sequence-determining information may be obtained by MIKES, CAD, and linked-field scanning methods (discussed later in this chapter).

6.5.1 Basic principles of field desorption mass spectrometry

Ionization of a molecule by FD processes depends on the presence of a high electric field gradient, on the order of one million eV cm^{-1} or one volt angstrom^{-1} (68b). This electric field gradient is sufficiently high so that a molecule in the vicinity of, and therefore experiencing that electric field gradient, loses or gains an electron to produce a positive or negative ion, respectively. That ion is accelerated out of the ion source and into the mass analyzer. Ion-molecule reactions generally occur on the emitter surface to produce, not the molecular ion, but rather an $(M + H)^+$ ion current and, at higher emitter heating currents, species such as (M+K) and (M+Na) anions and cations.

The FD emitter is generally a ten micron diameter tungsten wire covered with a dentritic surface (Fig. 6.8) which is produced under the influence of

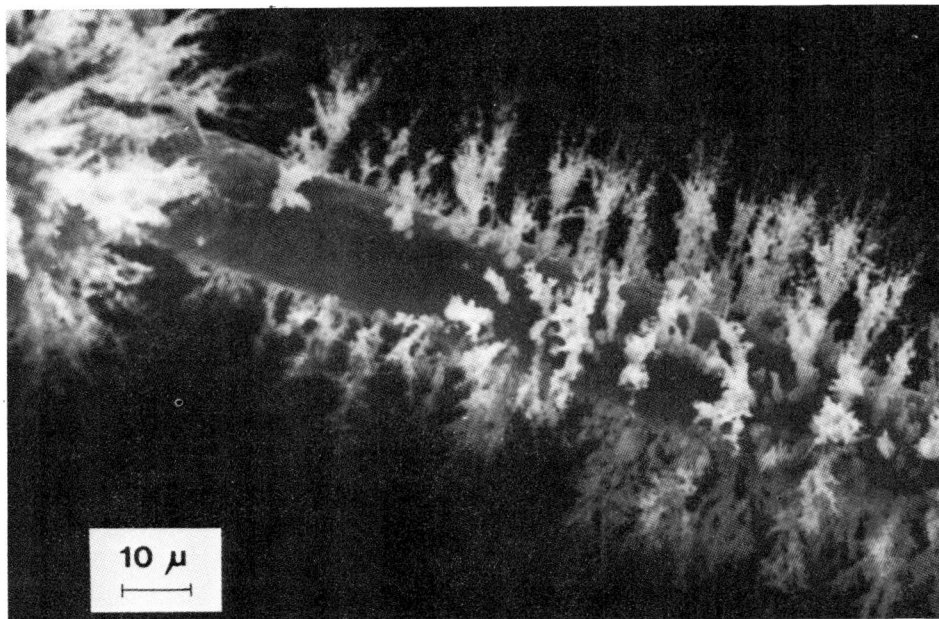

Fig. 6.8.   Scanning electron micrograph of an FD emitter.

a combination of several experimental parameters: an appropriate pressure
of an organic compound such as benzonitrile, high temperature, and a high
electric field. The length of the dendrites produced on an emitter surface
extends to approximately 30 microns.   At a sufficiently high level of applied
voltage, the radii of curvature of the outer tips of these dendrites produces
the required high electric field gradient.

An electric current (0-30 mA) passing through the emitter facilitates
melting and movement of peptide to the outermost tips of the emitter.   A
typical emitter heating current for FD-induced ionization of a peptide is
approximately 15 mA (68b).   Once a peptide molecule migrates to the emitter
tip, the high electric field gradient effects ionization, and positive or
negative ions are formed.   The resulting ion now resides on an
electrically-charged surface of several keV and within picoseconds, the ions
of like charge desorb from that emitter surface. This procedure of ionization
and desorption transfers little internal energy to the molecule, and minimal
(if any) fragmentation results.

Several methods can be used to deposit the peptide onto the FD emitter surface. In one method, a concentrated (one M) solution of the peptide is prepared in a tube which has a special bottom, and the emitter is dipped into the peptide solution and removed quickly. The peptide solution adheres to the carbonaceous emitter surface and when the solvent dries, the peptide molecules coat the emitter surface. In a second method, a drop of peptide solution ($\mu g\ \mu l^{-1}$) is placed between two emitters. When the two emitters are carefully separated, the drop will adhere to only one emitter (69). In a third method, controlled deposition of a peptide solution is produced by using a microsyringe with the aid of a stereomicroscope and micromanipulator ensemble (70; See Figure 6.9). Lastly, electrospray loading is effected by using a microsyringe at a high positive voltage (4kV) at a distance (2-5 mm) from the emitter where, under optimal conditions, 99% of sample is transferred from the syringe to the emitter (71).

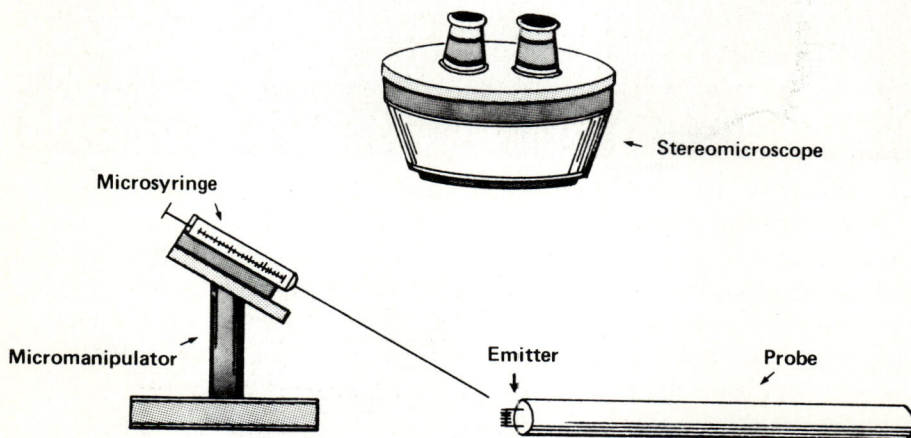

Fig. 6.9. Microsyringe-microscope-micromanipulator ensemble.

A review of the MS of nonvolatile and thermally unstable molecules (72) collects the various instrumental procedures currently available for ionization and analysis of that type of compound. The available ionization techniques include FD MS, fission fragment-induced desorption mass spectrometry, electrohydrodynamic ionization mass spectrometry, secondary ion mass spectrometry/fast atom bombardment (SIMS/FAB), laser desorption mass spectrometry, CI (in-beam desorption), and EI (in-beam, rapid heating, desorption, flash desorption). That review notes that every one of the techniques for ionization of nonvolatile and thermally unstable compounds yields mass spectra which exhibit only protonated and/or cationized molecular ions, or corresponding negative ions.

6.5.2 Types of field desorption mass spectra of peptides

Figure 6.7 shows an example of an FD mass spectrum of the polar hexapeptide KETYSK demonstrating an abundant $(M+H)^+$ and associated naturally-occurring isotope peaks, but no peptide bond fragmentation.

FD mass spectra are generally simple and, while this spectral simplicity offers significant advantages for analytical measurement (see Chapter 7), it is difficult to establish that the peak observed corresponds to the $(M+H)^+$ molecular ion of the peptide. Mass assignment may be either achieved or confirmed by using specialized experimental techniques. For example, an excess of a cation (sodium, potassium, lithium, cesium, rubidium) added to the peptide solution will shift the $(M+H)^+$ ion of a compound an appropriate number of mass units: $(M+Na)^+$, $(M+K)^+$, $(M+Li)^+$, etc.

FD MS has been utilized in studies of peptides, peptide antibiotics, neuropeptides, and glutathione peptides.

Several peptides were analyzed by FD MS (73). A comparison was made for three different ionization techniques (EI, FI, FD) for the acetylated, permethylated tripeptide Val-Gly-Leu. The EI mass spectrum showed molecular weight ions of low abundance but many fragment ions; the FI mass spectrum showed an abundant molecular ion and few fragment ions, while the FD mass spectrum showed only an abundant molecular ion peak. A chemically derivatized dipeptide (Z-Glu-Tyr) showed ions of low abundance in the molecular ion region in the FI mass spectrum, but both an abundant molecular ion and $(M+H)^+$ under FD conditions. The FD mass spectrum of a derivatized tetrapeptide containing two underivatized arginine residues (Ac-Gly-Arg-Arg-Gly-O-Me) indicated a very abundant molecular ion and an ion corresponding to peptide bond cleavage between the two arginine residues. The FI mass spectrum of the pentapeptide Phe-Asp-Ala-Ser-Val contained no molecular ion information, but an abundant $(M+H)^+$ occurred in the FD mass spectrum, in addition to several ions caused by peptide bond cleavages. This spectrum represents one of the first demonstrations that chemically underivatized oligopeptides could be analyzed by the FD technique. In all of these cases, the best anode temperature of the FD emitter is that temperature at which $(M+H)^+$ information is maximum, and fragmentation is minimum. That temperature is also a function of the condition of the emitter.

A sufficient amount of amino acid sequence-determining ion current was present in the FD mass spectra of the peptides Pro-Leu-Gly-amide,

CBZ-Gly-Pro-Leu-Gly-Pro, and the nonapeptide bradykinin Arg-Pro-Pro-Gly-Phe-Ser-Pro-Phe-Arg for amino acid sequence assignment (74). The FD mass spectrum of bradykinin indicated an $(M+H)^+$ of low abundance at 1,060 mass units and extensive peptide bond cleavage occurred to produce overlapping fragments. The 29 amino acid peptide glucagon was studied by FD MS (75). The peak of highest mass found in the mass spectrum corresponded to the N-terminal octadecapeptide when the sample was heated at a rate of one mA min^{-1} to avoid polymerization effects. A tetrapeptide amide (Trp-Met-Asp-Phe-NH$_2$) relating to the C-terminal amino acid sequence of human gastrin was studied with corresponding glutathione conjugates (76). Three basic hexapeptides containing two adjacent arginine residues were investigated with FD MS (77). Two fragmentation patterns specific to arginyl residues were noted in these FD MS mass spectra.

"Microderivatization" of peptides to be studied by FD MS was accomplished by using a mixture of methylation and methanolysis (78). Several structural features were determined including methods to differentiate $M^{+\cdot}$ from $(M+H)^+$, the number of carboxyl groups, identification of an N-terminal pyroglutamyl residue, and some amino acid sequence-determining information. A family of five synthetic tripeptides relating to pyroglutamyl-containing brain peptides was studied by utilizing FD MS and needleless emitters (79). A very small amount (50 pg of peptide) was placed onto the emitter. The authors conclude that needleless emitters generate sufficiently high sensitivity to determine the molecular weight of biologically important oligopeptides isolated at the submicrogram level but that further structural information is not available from the mass spectrum which is obtained on that amount. The oxidatively-roughened needleless wire emitters were used instead of benzonitrile-activated carbon emitters, where the former emitters were found to enhance ion formation by cation attachment.

6.5.3 Field desorption ionization mechanisms

Two groups have investigated the basic mechanisms which are involved in FD. One theory is prompted by the finding of ions at low electric fields, a fact which suggests that the electric field is of little importance in ion production (80). Data interpretation by these authors implies strongly the action of a field-free, temperature-dependent volatilization process which involves chemical attachment reactions in a thin fluidized system. These authors suggest a four-stage model to explain ionization in FD: solid state; semi-fluid with solute mobility; semi-fluid with small ion mobility; and fluid lattice with both thermal ionization and lattice ion mobility. Their hypothesis states that the electric field merely serves as a vehicle to facilitate removal of ions once they are formed by chemical reactions in the thin film.

On the other hand, a second group of investigators (81) discusses theoretical considerations of the FD process and their interpretation of several experimental facts (average electric field strength, field enhancement factor, supply mechanism, ion intensity versus field strength, neutrals, interface effects) disputes the field-independent thermal desorption mechanisms of ions.

The basic mechanisms for ionization of molecules deposited on a field desorption emitter has been studied (82). Both smooth metal emitters as well as dentrite-covered FD emitters were discussed.

The electrodhydrodynamic effects on smooth FD emitters were discussed, where the following mechanisms for the formation of gaseous ions are distinguished.

At the tip of a field-enhancing microneedle on an activated emitter, ions are formed by FI, or, in other words, field-induced electron tunneling reactions. The sample molecules are supplied to the tips, either from the gas phase or via surface diffusion to produce $M^{+\cdot}$, $(M+H)^{+}$, and M^{++}. The appearance of $M^{+\cdot}$ ions in FD mass spectra always provides strong evidence for ion formation by FI.

Field-induced desorption of ions from electrolytic solutions or salt layers is the most important mechanism for FD. Preformed solvated or electrochemically-formed ions are extracted from the condensed phase via charging of the sample layer surface which is exposed to the high field, where that charging depends upon thermally activated ionic conductivity of the sample layer. Typical ions are $(M+H)^{+}$ and $(M + alkali)^{+}$ ions. The appearance of the latter ions in FD mass spectra provides unambiguous evidence for the production of gaseous ions via a field-induced desolvation mechanism. Quasimolecular ions of so-called non-volatile compounds can be formed only via this mechanism.

Thermionic emission of metal ions is observed at emitter temperatures greater than 800° K.

It is emphasized that the extraction of ions from condensed sample layers is the only important mechanism in FDMS for the generation of gaseous quasimolecular ions from thermally labile compounds. In an effort to elucidate the mechanism by which ions are extracted from a deposited condensed sample layer, optical microscopy in conjunction with FD has been utilized to study the behavior of such a layer (82). For organic compounds such as sucrose, no flux of liquid (charged sample layer) from surface areas of low to those of high field strengths was observed. That result does not

exclude the flux even of the charged sample layer in a region close to the
emitter tip, as it was observed in the study of metals. An important result
was the observation that most of the sample is lost from field anodes by the
formation of charged droplets. In a sucrose solution, the effect of temperature
on a sample droplet is to decrease the surface tension of the liquid slightly,
and to reduce the viscosity of the sample strongly. The latter reduction
leads to increased mobility of the solvated ions, which then causes a charging
of the droplets in the high electric field. For sucrose solutions, the stress
which is caused by the electric field first forms bubbles due to the evaporation
of solvent molecules within cavities. Rupture of the bubbles then leads to the
formation of jets which are composed of small charged particles. During the
destruction of the bubbles, lamellae and points, strong oscillations of the
projections are observed which are caused by the emission of charged particles.
Molecular ions are detected only after most of the sample is lost into the gas
phase and the disintegration processes have become negligible. The final state
of the sample (which was not previously observed with experiments at a lower
resolving power of the microscrope) are presented.

For tartaric acid, application of a high field alone (without heating)
leads to the formation of projections of various kinds and jets. Very fine
points are considered as emission centers for $(M+H)^+$ ions.

The response of charged surfaces of organic liquids to the stresses
which are exerted by a strong electric field (electrospray phenomenon) has
been investigated experimentally and theoretically for various geometries
and for liquids. The micrographs presented show that the sequence of
disintegration phenomena continues to proceed in microscopic volumes of
liquid. The rate of disintegration is dependent mainly on the viscosity and
electrical conductivity of the sample. Since an increase of emitter temperature
leads to a decrease of the viscosity of the liquid sample, temperature is
the most critical parameter for the observed electrohydrodynamic effects.

The field strength applied to a sample must be sufficiently high to
create electrohydrodynamic instabilities. For small sample droplets deposited
onto a 10 micron wire, experiments show that the initiation of disintegration
of the sample at elevated temperatures almost automatically results in the
formation of very small field-enhancing tips and fibers from which molecular
ions are emitted.

Other experiments reveal that molecular ion emission occurs from an
amorphous nearly solid or glassy state from the sample which has a loose
structure with many cavities (83). This state is formed via the growth of
field-enhancing protuberances from a viscous, liquid state and the evaporation

of solvent molecules from the surface of the protrusions. The temperature change of the molecular ion emission is significantly below the melting point of sucrose. Observed changes in the ion energy distribution and intensity fluctuations are consistent with ion emission from transient glassy protuberances of low ionic conductivity. In contrast to a previous study, the ion emission from the projections observed in the microscrope was possible. This direct correlation is observed between optical micrographs and molecular ion emissions and reveals that the ions originate from an amorphous nearly solid or glass state of the protrusions formed. The protuberances are not well-shaped as would be expected for a viscous liquid. They were found to stay in vacuum, even after switching off the electric field, but collapse and turn into a liquid state again when exposed to air; most probably by the re-uptake of water from the air. The development of these protuberances from a viscous-aqueous solution of sucrose is easily explained by the evaporation of solvent molecules from growing projections. The authors reach the conclusion that optimal conditions from molecular ion emissions are achieved by applying a slow rate of emitter temperature increase, and by limiting the emission current to a pre-selected low value.

In a study (84) on the field desorption of sucrose, several sequential steps are formally distinguished:

In the first stage, the applied electric field causes a charging of the layer. The resulting field stress gives rise to electrohydrodynamic instabilities and a disintegration of the layer, and as a result field-enhancing protuberances are formed. Heating of the sample is needed to decrease the viscosity and, hence, to raise the rate of disintegration.

In the second stage, during the growth of protrusions having a large surface-to-volume ratio, an amorphic, or nearly solid, glassy state of sucrose is formed via evaporation of solvent molecules. Of course, higher emitter temperatures are in favor of solvent evaporation while, on the other hand, they also give rise to an increased rate of disintegration of the sample via droplet or cluster ion emissions. A further effect of heating on growing projections which turn into a glassy state is a decrease in ion conductivity.

In the third stage, the field strength on the surface of a growing protrusion becomes high enough for the extraction of molecular ions. At least the surface region of the protuberances must be in an amorphic solid state to account for the fact that typically only monomolecular ions are formed, but no higher cluster ions. In contrast, a liquid state of the sample, or even a high mobility of surface molecules, gives rise to the emission of a broad distribution of cluster ions as is well known from electrohydrodynamic

ionization mass spectrometry, from studies of low molecular field
ionization of alcohols, and from ion evaporation from liquid droplets.

The action of an electric field on alkyl ions results in the slow
rearrangement of the surface on the molecular scale and, hence, on the
extraction of molecular ions via successive ruptures of intermolecular
hydrogen bonds. The "dissolvation" of even a single ion involves the
rearrangement of a number of molecules. Such a cooperative mechanism of
ion extraction from the "molecularly rough" surface is required to overcome
solvation energies in the range of some electron volts by displacement of
charges in fields of less than 3 volts per nanometer (below the onset field
strength of field ionization). It is obvious that the dissolvation mechanism
is, in all parts, a thermally activated process.

In the case of activated emitters, sample layers may be exposed to
field strengths which are high enough to allow the extraction of ions without
the preceding formation of field-enhancing projections from a liquid surface
of the sample. However, thermal heating is still required for this process,
less for the evaporation of solvent molecules than for producing a rough surface
structure and for ion extraction. Under these conditions of high field strength,
the adhesion of the sample to the microneedles must be sufficiently strong. It
is of interest that field ion emission from charged droplets in the "thermospray"
method is independent of adhesion of a sample to a substrate and, hence, avoids
that problem.

6.5.4 New instruments constructed to attain higher masses

Whenever it was realized that FD could provide ions at mass values much
higher than it had previously been possible, investigators began to fabricate
newer MS instruments to analyze those higher masses (85).

Larger mass spectrometer instruments have been designed and fabricated
(86,87) using a small magnetic sector angle (55°). The mass range of a mass
spectrometer operating at a fixed accelerating voltage is determined by both
the magnetic field intensity and its radius. The weight of the laminated (245
layers) magnet (1.4 Tesla) in this "grand scale" instrument is 7 tons, has a
radius of 780 mm, and permits analysis of masses up to 8,250 a.m.u. A
theoretical limiting resolution of 25,000 and a mass range of 20,000 is
calculated. Recent advances in magnet design now permit analysis at fast
scanning (100 msec decade $^{-1}$) and high mass range (20,000 a.m.u.)
specifications.

The specifications of an MS instrument which is amenable to high mass
analysis following rapid and facile production of the mass spectrum of large
molecules at full sensitivity were developed (88). These specifications

extend the mass range up to mass 3,000 by using a high saturation alloy of cobalt-iron-vanadium (permendur). This research resulted in the production of the first commercial high field magnet on a mass spectrometer. The calibration curve of this magnet in the high field mode demonstrates extension of the mass range up to mass 3,200 at full accelerating voltage (8,000 volts), where saturation was noted at mass 2,700.

Currently, several developments are being incorporated into commercially available mass spectrometers to achieve increased sensitivity at high mass. Some operating specifications demonstrate ions at 20,000 mass units for a CsI marker compound. Extended sectors, laminated magnets, controlled inhomogeneities in the magnetic sectors, "rolling" magnetic field sectors, specifically-shaped magnetic field pole pieces, and other developments are now being incorporated into the instrument fabrication process. Higher masses can also be observed readily by decreasing the accelerating voltage, although that method of increasing the mass range is done at the sacrifice of sensitivity.

6.5.5   Field desorption mass spectra of larger peptides

FD mass spectra were obtained from trypsin treatment of the alpha, beta, gamma, and delta chains of normal human hemoglobin and the beta chain of sickle hemoglobin (89). Almost every one of the expected $(M+H)^+$ ions of the proteolytically-produced peptide fragments and corresponding doubly-charged peaks were observed; one ion was observed at m/z 2,955. The molecular ions of these tryptic digest peptides corroborate the theoretical $(M+H)^+$ ion masses of expected products.

A combination of FD MS and Edman degradation was utilized to determine the amino acid sequence of an unknown peptide containing 55 amino acid residues and which is derived from the N-terminal cyanogen bromide fragment of the Streptomyces erythraeus lysozyme (90, 91).

FD MS permits study of those compounds which have molecular weights in a mass range higher than that amenable to previous EI and CI methods. Research done at higher masses requires the use of high molecular weight reference compounds such as hexakis-(multifluoroalkoxy)-cyclotriphosphazenes (92). Multivalent cations have also been effectively used to extend the practical mass scale in FD MS (93). For example, a barium salt was used to produce $[M + Ba]^{++}$ ions of sucrose, raffinose, and stachyose.

6.6   QUANTITATIVE FIELD DESORPTION MASS SPECTROMETRY

FD MS has been utilized in several laboratories for quantitative analysis of biologically important compounds. Advantages of and sources of error for different internal standards, experimental methodology, and examples of applications have been discussed (94). Ion recording methods include photographic plates, repetitive scanning, selected ion monitoring (SIM), and the double detector technique. The utilization of structural homologs and stable isotope-incorporated compounds for internal standards improves precision, accuracy, and sensitivity. In both quantitative analysis and biochemical/biomedical research, the outstanding features that MS detection can offer to the analyst include high sensitivity and high specificity. When compared to EI and CI methods, relatively low and often strongly fluctuating ion currents are generated in FD MS, a fact which limits sensitivity and precision in quantitative analysis. On the other hand, the commonly observed high relative intensities of $(M+H)^+$ ions or cationized molecules found in FD mass spectra are favorable for two reasons:

(i) optimum sensitivity can be obtained, since the molecular ion group frequently carries 100% of the total ion current, and

(ii) since predominantly only one ion species is produced from each component of the mixture, the possibility for interference by impurities is minimized. Using the SIM mode for the determination of cesium in various solvents, body fluids, and environmental samples, between one nanogram and one picogram of cesium is detected with a precision of ±10% and an accuracy of ±20%.

The main advantage of the photographic recording of FD mass spectra is the simultaneous registration of all ions within a wide mass range (42, 95, 96). The sensitivity limit of ion detection can be improved considerably by the use of gelation-free photoplates, where the silver bromide crystals are located closer to the surface compared to normal gelatinous plates (97). It is noted that the mass spectrometer must be scanned for electric detection and the greater part of the measuring time is not used to record the ions, but rather is used to scan between ion masses, especially in a high resolution mode. On the other hand, quantitative determinations with a photographic emulsion are more tedious because calibration of the ion response of the photoplate is required for each analytical measurement. Because of the limited dynamic range of sensitivity of a photoplate, multiple exposure levels are required. A multichannel analyser was used for integrating electric recordings in a study of phenylthiohydantoin amino acids. Incorporating a stable isotope such as deuterium played a significant role in preparation of internal standards. Compared to the C-H bond, the deuterium-carbon bond is stronger and the corresponding molecular volume is smaller, two structural parameters which result in a more lipophilic compound after deuterium labeling. On the other hand, the stable isotope-labeled compounds containing

^{13}C, ^{15}N, and ^{18}O are more suited for production of internal standards. A mass difference of at least three mass units between the $(M+H)^+$ ions of the sample and the internal standard is required, because this degree of mass separation considerably simplifies data evaluation. One reason for this simplification in data analysis is that the superimposition of the natural ^{13}C signals of the sample with those of the stable isotope-labeled internal standard can be neglected.

Orally administered berberine chloride was analyzed in human urine by utilizing FD MS in the SIM mode where a deuterated analog was used as an internal standard (98). Berberine chloride was determined at an endogenous concentration level of ten ng ml^{-1} human urine. The quatenary ammonium ion base peak which was monitored arises via loss of a chlorine atom from the molecule. A sample-loading device was fabricated from a tungsten needle where a drop of the sample solution formed by a microsyringe was transferred to the device and loaded onto the FD emitter. While the lifetime of one emitter was examined in a study where one emitter was used repeatedly up to 90 times, the authors found that each emitter could be used more than 40 times. The detection limit of this method was two picograms, with a signal-to-noise ratio greater than 3:1.

Biologically important oligopeptides, especially the opioid penta-peptides ME and LE, have been the object of measurement utilizing FD MS. Details of the following analytical techniques will be discussed more completely in Chapter 7. An off-line combination of RP-HPLC and FD MS is utilized for a study of hypothalamic oligopeptides (70). The concentration of endogenous ME in canine tooth pulp was studied with this methodology using the higher homolog compound ^{2}ala-leucine enkephalin as internal standard (99). The combination of FD MS-CAD-linked field (B/E) scanning MS was the analytical technique chosen for analysis of underivatized oligopeptides. This instrumental combination of techniques was utilized for a study of the peptides LE, Phe_2, and Phe_4 (100). Solutions of synthetic LE methyl ester were quantified down to 35 ng (52 pmol) by means of FD MS in the SIM mode (101). Two enkephalins, LE and ME, were measured in one experiment utilizing the two respective stable isotope-labeled internal standards, $^{18}O_2$-leucine enkephalin and $^{18}O_2$- methionine enkephalin. FD MS was utilized to produce $(M+H)^+$ ions and the two enkephalins were measured in canine caudate nucleus tissue extracts where ME (985 ng g^{-1} tissue) and LE (162 ng g^{-1} tissue) were measured (102).

6.6.1 Experimental limitations of FD-MS

The practical experimental limitations of FD MS include both the critical emitter preparation step and the short-lasting, rapidly fluctuating $(M+H)^+$ ion current. While endogenous peptide measurements in biologic extracts were made originally with FD MS measurements of $(M+H)^+$ ions, these experimental limitations hampered a more extensive use of the FD MS measurement procedure. A third limitation is unavoidable emitter breakage experienced in about 30% of the experiments.

6.7 FAST ATOM BOMBARDMENT MASS SPECTROMETRIC STUDIES OF PEPTIDES

While FD MS permitted an important advancement in MS studies of biologically important peptides, another more significant advance in peptide chemistry soon followed with the development of FAB MS.

6.7.1 Basic principles of fast atom bombardment mass spectrometry

Secondary ion mass spectrometry (SIMS) has been utilized for years, where an ion of high translational energy is directed at a sample target and secondary ions which sputter from that target material are mass-analyzed (103-105). Amino acids, steroids, and folic acid derivatives were studied.

While organic and bioorganic compounds were studied by SIMS, a significant advancement was achieved simply when a liquid matrix was used to minimize sample destruction and stabilize the production of ions. The liquid matrix effectively regenerates a fresh molecular surface for ion/atom bombardment, while also reducing the degree of sample burning (106-112).

The basic details of FAB are represented in Figure 6.10. A noble gas (commonly Ar° or Xe°) is ionized, accelerated to high translational energy (8-10 keV), neutralized, and exits the FAB gun to impact onto a probe tip. The probe tip is appropriately beveled and contains sample dissolved in a non-volatile viscous matrix. The term "angle of-incidence" is used extensively in the FAB MS literature and it is well to remember that this term has a definite physical meaning (113) to describe the interaction of impinging radiation with surfaces and is defined as the angle between the direction of incidence and the normal to the surface at the point of incidence (114, 115).

The most commonly used FAB matrix is glycerol. Thioglycerol, GC coating material, plus other appropriate matrices may be used. The momentum of the impacting fast atom is transferred to the molecule, an $(M+H)^+$ ion is formed by a process involving the energy from momentum transfer, and the $(M+H)^+$ is ejected from the probe surface and enters into the analyzer section of the mass spectrometer.

A description has been published outlining the modifications of a
conventional MS ion source housing to accept a commercially available neutral
ion gun (116, 117). The evaluation of the various experimental parameters
which effect the ionizing efficiency and sample ion yield have been described,
and the majority of this next section describes the research data published by
that group (118). Probe tip alignment is achieved visually by using the light
emitted by the FAB gun (blue for xenon, purple for argon, and red for
neon). The standard procedure for FAB analysis is to coat a stainless steel
tip with 50% (v/v) nitric acid for a few seconds, rinse it with distilled water
and nanograde methanol, and allow it to dry. Samples (usually at a
concentration of $\mu g \ \mu l^{-1}$) are applied to the probe tip and dried. Glycerol (300
nl) is spread on the surface of the probe tip so that a thin, uniform film is
visible to the unaided eye. When more than one microliter of sample is applied
to the sample tip, a hemispheric shape is produced which alters the
angle-of-incidence of the impacting fast atoms and causes a decrease in the
sensitivity. An adaptor is utilized to mount the atom gun onto the ion source
housing and to keep the FAB gun out of the pathway used to remove the ion
source. In an attempt to lower the operating pressure of the ion source
without decreasing the efficiency of the FAB gun, a gun exit hole of 0.5 mm
is used. The commercially available FAB gun (Ion Tech B12N, Teddington,
England) is a saddle-field source (116, 117) which generates diametrically-
opposed ion beams of equal intensity, one of which strikes a beam monitor
plate and indicates ion current. The other portion of the ion beam travels
through a flight tube, where it is neutralized by a dense cloud of secondary

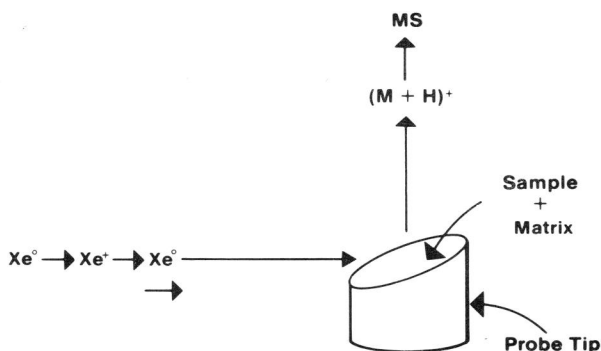

Fig.6.10. Basic principles of FAB mass spectrometry.

electrons produced when the ions strike the sides of the flight tube. The pathway to the monitoring ion beam is blocked by a solid piece of aluminum to force all the gas to flow into the front section of the gun to permit lower ion source operating pressures (118).

The target material for the FAB gun was also studied (118). Copper, 303 stainless steel, and polyimide-coated stainless steel were investigated. Copper performs in a satisfactory manner, except occassionally when it forms cluster ions with the sample or matrix. Those cluster ions may complicate the mass spectrum of a compound under investigation. For example, copper-containing ions are found at the accurate masses 161.8586 ($C_3{}^{63}Cu_2$), 208.7753 ($O^{63}Cu^{65}Cu_2$), 214.7810 ($NaH^{63}Cu_2{}^{65}Cu$), among many others. Furthermore, the copper tips must be replaced frequently because the dilute nitric acid used for cleaning tips between samples significantly erodes the copper. On the other hand, 303 stainless steel is not etched with 50% nitric acid and is found to have no sample memory. While the wettability of the 303 stainless steel is not as good as that of copper, it is still adequate. Polyimide-coated tips were prepared to study the potential effects of surface-charging of the sample. Surface-charging effects, however, do not play a role.

To determine the optimum angle-of-incidence for FAB MS, seven probe tips were machined which had incidence angles ranging from 30° to 90°, in 10° increments (118). In each case, the same experimental conditions of ion current, pressure, temperature, and sample concentration were used; the only variable was the angle-of-incidence to the particle beam. A plateau is reached at 60° for the relative intensity of the $(M+H)^+$ ion of a peptide. The angle-of-incidence was also investigated with relation to the effect on the extent of decrease of ion current when changing from low resolution (one part in 1,000) to high resolution (one part in 10,000). No significant change in sensitivity is observed for the various angles for a probe tip of 2.5 mm diameter, or 4.9 mm² area. This result of equivalent sensitivity is not unexpected, because the diameter of the probe tip and therefore, the area over which the sample is spread, is larger than the slit width (one mm) in the extraction electrode of the Finnigan MAT 731 mass spectrometer, which is the first ion lens in the combination source. Smaller diameter tips (one mm.) were also used, but a drawback is that their sample capacity is six-fold less than for the 2.5 mm tips (4.9 mm² surface area vs. 0.8 mm²).

A rotatable probe was developed for accurate mass measurement (within ± one millimass unit) during FAB MS (119).

The gas used to ionize the sample in FAB MS plays the major role in determining the total ion current produced from that sample (118). The abundance of the $(M+H)^+$ ion of a peptide is plotted against the mass of the bombarding gas atom. While the mechanism of ionization in FAB MS is still not well-defined, it has been shown that, for monoatomic gases, the ionization efficiency of a specific sample is directly proportional to the atomic mass of the incident, neutral, fast-moving particles. Advantage is taken of this proportionality between ionization efficiency and mass of incident particles. In one study, organic samples were studied using mercury as a projectile (120) and in another study by using molecules derived from a diffusion pump fluid trimethylpentaphenyl trisiloxane (DC 705) (121) to take advantage of projectiles of large mass.

In these studies, sensitivity was not defined in the conventional expression (coulombs of ion current produced per microgram of sample) but rather in the operational sense as the minimum amount of sample required to provide a signal for the $(M+H)^+$ ion of the compound that is clearly distinguishable from the background signal, where the latter ion current is mainly that chemical noise generally due to the matrix (118). The ratio of the signal versus background noise in the positive ion mode is greater than 40:1, 17:1, and 25:1 for the peptides LE, $[Sar^1, Ala^8]$angiotensin II, and bradykinin, respectively, at a sample concentration of one mg ml^{-1} each.

Two major problems were found in sample preparation for FAB: sample dissolution in a matrix-compatible solvent and presence of salts (118). Solubility problems can be partially overcome by using acids for basic samples and bases for acidic samples to facilitate their dissolution prior to adding the matrix. Often, either sonication of the solution for a short time or allowing a sample to stand overnight will greatly increase the ion abundance, due presumably to an increase in sample concentration. Another problem can be the apparent suppression of the FAB MS by salts. This suppression can be eliminated readily by utilizing either a Sep-Pak or an HPLC purification step (see Chapter 4) prior to FAB MS. A list of matrix-associated ions observed in a FAB mass spectrum includes components glycerol, sodium, potassium, chloride, ammonium, formic acid, guanidine hydrochloride, phosphate, diaminobutane, oxalic acid, acetic acid, hydrochloric acid, heptafluorobutyric acid, and ammonium trifluoroacetate.

6.7.2 Types of fast atom bombardment mass spectra of peptides

In general, the abundant total ion current produced by FAB of a peptide is comprised of multiple ionic species including both protonated and cationized molecular ion currents due to both peptide and matrix. Because of the impacting fast atom beam and the corresponding large amount of translational energy which is transferred to the peptide molecule, fragment ions are also generally produced. A disadvantage of FAB MS studies with a liquid matrix is that the matrix also produces an intense mass spectrum which overlays the peptide mass spectrum. This overlaying phenomenon is similar to using perfluorokerosene as a mass marker in EI MS. A continuum of peaks at virtually every mass is observed in the FAB MS of glycerol, with major peaks occurring periodically at (92 x n + H) mass units. The mass spectrum in Figure 6.11 illustrates this repetitive nature of glycerol polymer ions which are formed during the FAB of neat glycerol. The $(M+H)^+$ ion of glycerol occurs at 92 x 1 +H = 93 mass units, and this ion adds from one to nine (and more) molecules of glycerol to produce the ions observed at masses 185, 277, 369, 461, 553, 645, 737, 829, and 921, respectively.

It is important to unambiguously elucidate and verify the genesis of each one of the ions found in Figure 6.11. This elucidation is important for two reasons. On one hand, the glycerol FAB mass spectrum may be subtracted from a composite spectrum of sample and glycerol, whenever that subtraction procedure is appropriate. On the other hand, if a glycerol polymer ion persistently interferes with a critical sample ion, then d_5-glycerol can be used as an alternate matrix to shift the matrix ion away from the sample ion.

Neat $HO-CD_2-CD(OH)-CD_2OH$ can also be used as the matrix, and the corresponding mass spectrum is shown in Figure 6.12. The first glycerol monomeric ion observed in Figure 6.11, mass 93, shifts the appropriate five mass units in the mass spectrum (Figure 6.12) of d_5-glycerol. Each succeeding multimer shifts (5 x n) mass units, or 10, 15, 20, 25, 30, 35, 40, 45, and 50 mass units corresponding to masses 195, 292, 389, 486, 583, 680, 777, 874, and 971, respectively, in Figure 6.12.

The bombarding particle beam used in FAB MS may possibly effect chemical changes in the sample and produce a mass spectrum not characteristic of the sample under investigation. One must also keep in mind that radiation damage observed for glycerol may possibly occur for molecules dissolved in glycerol (122). Good laboratory practice indicates that FAB mass spectra should be obtained as soon as possible after probe insertion in an effort to minimize the amount of FAB (and potential damage) to both the glycerol matrix and the solute molecules.

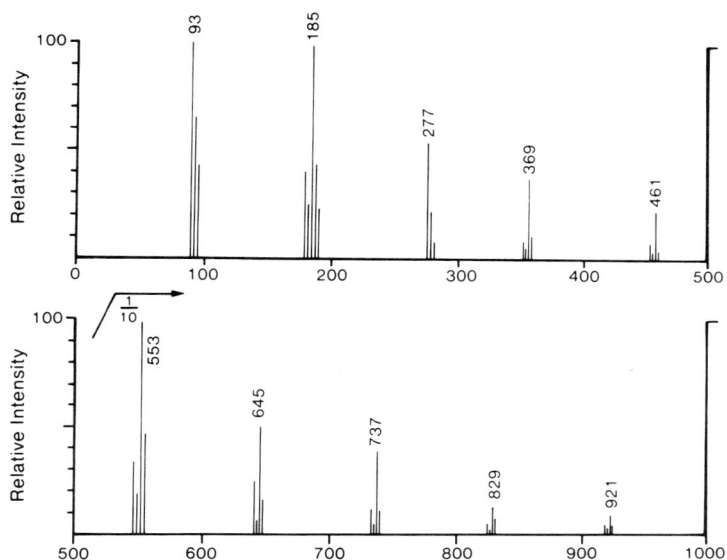

Fig. 6.11. FAB mass spectrum of glycerol.

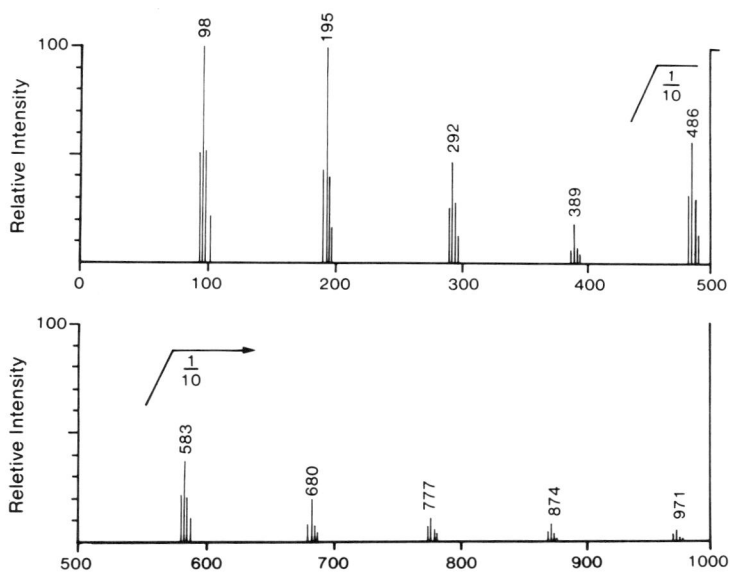

Fig. 6.12. FAB mass spectrum of d_5-glycerol.

6.7.3 Fast atom bombardment mass spectrometry of peptides of high molecular weight

Proinsulin, a compound having an elemental composition corresponding to $C_{410} H_{638} N_{114} O_{127} S_6$ and a molecular weight of 9,382.529 mass units, has been studied with FAB MS (112). The FAB mass spectrum of this peptide molecule which has a relatively high molecular weight is obtained at a reduced ion accelerating voltage and with a fast atom beam "current" of one to two mA, where the peptides are dissolved in a mixture of alpha-monothioglycerol plus glacial acetic acid. An unresolved cluster of ions in the $(M+H)^+$ ion region is observed, but, due to high background noise, no amino acid sequence-determining ions are found. It is interesting to note that, at such a high mass range, the monoisotopic $(M+H)^+$ ion is only a minor component in the isotopic distribution. The presence of a large number of hydrogen atoms and other natural isotopes combine to shift the highest isotopic peak measured from the isotope profile which would be normally observed (123).

This type of mass spectral data of high mass bioorganic compounds demonstrates the fact that the experimental limitation to FAB MS studies is no longer primarily due to the ionization process, which has been the most significant restraint in the MS of high mass compounds up to now, but rather to characteristics of the instrumentation, especially the magnetic sector.

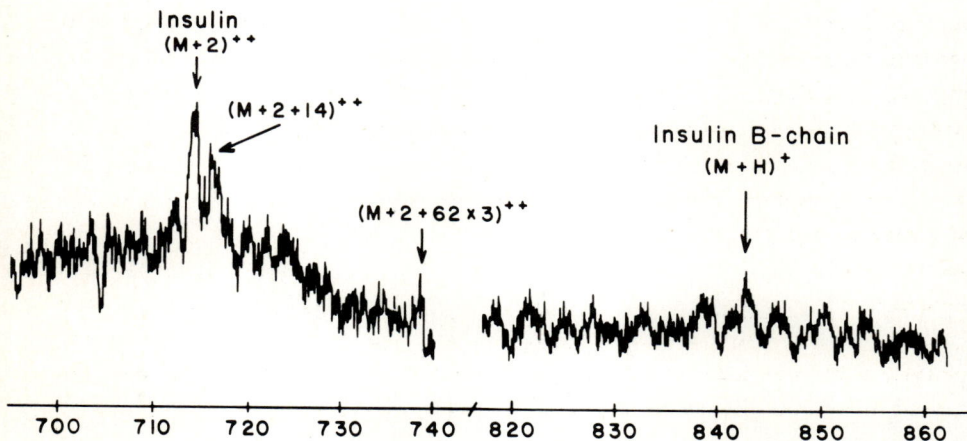

Fig. 6.13. FAB mass spectrum of insulin obtained on a Finnigan MAT 731 instrument (125).

Several laboratories have reported on the FAB MS of the intact insulin molecule containing both the A and B chains (111, 124, 125). This molecule has a molecular weight of 5,627 mass units. Figure 6.13 contains the FAB mass spectrum of insulin obtained on a Finnigan MAT 731 instrument, where a doubly-charged mass is observed at mass 703 (713 uncorrected) at 4kV acceleration and a mass marker indicating mass at one-half values. Therefore, the observed, corrected mass of 703 is multiplied by a factor of 8.

FAB MS is used in studies of glucagon, proinsulin, insulin, insulin A-chain, insulin B-chain, monellin, mellitin, bradykinin, fibrinopeptide, glycoproteins, peptide antibiotics, enkephalins, endorphins, substance P, angiotensin, and neurotensin, among others.

FAB MS is also utilized in both the positive and negative ion modes in a study of the opioid pentapeptides, ME and LE, where the negative ion mass spectra were noticeably less complex (126). Following development of FAB MS, a series of papers was published to illustrate the utility of this novel ionization technique for study of increasingly longer oligopeptides. Bradykinin, lysyl-bradykinin, and methionyl-lysyl-bradykinin were studied (107). Both positive and negative ions were discussed and amino acid sequence-determining ions were found. Ion transmission through the analyzer(s) of a mass spectrometer is independent of polarity, but when using an electron multiplier as an ion detector, there is a loss of detector response due to the reduced energy with which the negative ions impact on the negatively-charged first dynode of the detector. This difference in detector response leads to a reduction in available dynamic range when compared to the response range which is observed with positive ions. A limited discussion proposed quantitative analysis. It is significant to note that the biologic activity of the peptide before and after fast atom irradiation indicates no loss of biologic activity using the in vitro guinea pig ileum bioassay method. Both ^{5}Ile-angiotensin I and II were studied (108). Positive and negative ion mass spectra are produced and, in one case, a mass-analyzed ion kinetic energy spectrum (MIKES) of the $(M+H)^+$ ion is presented. In general, fragmentation is reduced in the negative ion mass spectra of these peptides. Smaller amounts of the two angiotensins were derivatized with a one-to-one mixture of d_3- and d_0-acetic acid anhydride to produce a doublet of ions differing by three mass units of all of the N-terminal-containing ions.

Larger peptides which have been studied include mellitin (MW=2,845), glucagon (3,841), and the B-chain of bovine insulin (3,494) (110). Mellitin is a constituent of bee venom and glucagon is an insulin antagonist. Using a normal ZAB 1-F reverse geometry mass spectrometer at full accelerating potential, an unresolved complex of ions is found in the molecular ion region of mellitin with an $(M+H)^+$ ion of 2,845.8. Another instrument outfitted with a high-field magnet with a mass range of 3,000 mass units at full accelerating potential demonstrates baseline resolution of the $(M+H)^+$ ion region. The positive ion mass spectrum of the B-chain of bovine insulin (disulfonic acid derivative) showed ions in the $(M+H)^+$ ion region as did the mass spectrum of glucagon. Masses were counted manually by comparison with the FAB MS of potassium iodide cluster ions $[K(KI)_n]^+$ and agreed to within one mass unit of the appropriate theoretical value. It is significant to note that, in the mass spectra of the peptides angiotensin, bradykinin, mellitin, glucagon, and insulin, ions were also observed 14 mass units higher. The source of those ions is not yet known, but may arise from the matrix.

$(M+H)^+$ ions produced from intact bovine insulin (5,730) and ACTH (4,538) are reported (111) in a paper which describes a useful mass range - defined traditionally as the range over which unit mass resolution is maintained.

A range of polar organic compounds was studied by FAB MS including organic salts, polar antibiotics, nucleoside phosphates, and underivatized peptides (127). Mass spectra with abundant $(M+H)^+$ ions are found for somatostatin, gastrin, and substance P. Pre-existing ions are proposed to pass into the gas phase via a momentum transfer mechanism, where that momentum derives from the inert fast-moving gas atom that impacts near, but not upon, the sample ion. Glycerol is normally used as the matrix, but when increased polarity is needed, tetragol $[HO(CH_2CH_2O)_4H]$ and teracol $[HO(CH_2CH_2CH_2CH_2O)_nH]$ are used. The authors note that there is an advantage in terms of high sensitivity in obtaining mass spectra of compounds with a net positive charge at pH 6.5 in the positive ion mode, and those with a net negative charge at pH 6.5, in the negative ion mode. For example, substance P yields an $(M+H)^+$ ion at 1,347 for one microgram (740 pmol) with a signal-to-noise ratio of 40:1.

A series of "difficult" peptides was studied with FAB MS and includes two peptides from cytochrome C-550 of the denitrifying aerobic bacterium Paracoccus and the second peptide was from the C-terminus of that protein (128). The third peptide was from the N-terminus portion of a dolphin cytochrome C.

A detailed FAB MS study is reported for peptides containing up to 21 amino acids. Positive and negative ion mass spectra are obtained with sample size ranging from two to 15 nmol of peptide (129). In favorable cases, molecular weight information is derived from 100 pmol of sample. Mass spectra are produced for substance P, human fibrinopeptide peptide A, gastrin-13, and various peptide enzymatic digests.

A combination of protease, carboxypeptidase digestion, and subtractive Edman degradation coupled with FAB MS is used to obtain amino acid sequence-determining information from substance P and an eledoisin-related peptide (130). Generally, 100 picomoles is sufficient for determining the molecular ion of a peptide in either the positive or negative ion mode. However, a larger amount of sample (1-5 nmol) generates some amino acid sequence-determining ions.

An adrenal zona glomerulosa-stimulating component of a posterior pituitary extract was characterized by FAB MS as bis-acetyl-^{1}ser-alpha-melanocyte stimulating hormone (MSH) (131). HPLC fractionation yields two major peaks of aldosterone-stimulating activity, where the second component is characterized as the bis-acetyl hormone. FAB MS of the biologically active material demonstrates an $(M+H)^+$ ion at mass 1,706. Chymotrypsin treatment, followed by EI MS of the N-terminal dipeptide, demonstrates the site of O-acetylation. Bioassays using rat adrenal glomerulosa show that the biological activities of bis-acetyl-^{1}seryl-alpha- MSH and alpha- MSH are virtually identical. The first HPLC peak is MSH.

Eleven zervamicin and two emerimicin peptide antibiotics were studied by FAB MS (132, 133).

Two peptides, intact monellin and insulin, which have molecular weights in the 6,000 dalton range, were studied by FAB MS (124). Monellin contains two subunits of molecular weights 5,248 and 5,831, respectively. One subunit is heterogeneous, where approximately 10% of those subunit chains contain an additional Phe residue at the N-terminus. Insulin was utilized as a test compound to optimize resolution and scan rate conditions of the instrument before studying monellin. The proteins were dissolved in a mixture of dimethylsulfoxide and 1M hydrochloric acid (3:1, v:v). Cesium iodide cluster ions were used for mass calibration at masses 5,069.3 ($Cs_{20}I_{19}$), 5,329.1 ($Cs_{21}I_{20}$), 5,588.9 ($Cs_{22}I_{21}$), 5,847.7 ($C_{23}I_{22}$), and 6,108.5 ($Cs_{24}I_{23}$). This study demonstrates that, for the first time, polar organic molecules up to a molecular weight of 6,000 daltons are rather readily ionized and detected using FAB MS and high-field magnet technology. The MS data are relatively difficult to obtain, except for the insulin, and require fine-tuning of the mass spectrometer and good basic instrument sensitivity. The accuracy of mass measurement at these relatively high masses is difficult to routinely estimate to within ±two mass units by current methods. While a mass assignment error of two units is extremely high compared to the experience gleaned from MS techniques used up to this point, in studies of protein and carbohydrate chemistry at such very high masses, this degree of mass determination accuracy is highly significant and demonstrably useful to readily distinguish one amino acid or monosaccharide residue from another.

A discussion of a variety of ionization techniques is presented (134) and includes FD, plasma desorption, SIMS, electrohydrodynamic ionization, laser desorption, thermal desorption, and FAB MS. A cyclic tetrapeptide was studied utilizing FAB MS (135).

6.8 OTHER FAB MS TECHNIQUES

The CAD mass spectrum of the FAB-produced $(M+H)^+$ ion of the heptapeptide Ala-Leu-Trp(For)-Asn-Phe-Arg-Ala yields a greater abundance of amino acid sequence-determining ions compared to the normal FAB B-scan mass spectrum (136). Both the metastable ion and CAD spectra are obtained by scanning the electric sector voltage at a rate of 16 volts $\sec^{-1}$. Mass spectra are obtained from estimated amounts (50-100 μg) and were recorded with a UV chart recorder. The mass spectra of all compounds show intense $(M+H)^+$ and $(M-H)^-$ ions in the positive and negative ionization modes, respectively. Both the intensities and the signal-to-background noise ratios of these molecular ion peaks are improved when the compounds are dissolved in thioglycerol instead of glycerol. While the metastable ion mass spectrum of the $(M+H)^+$ of the heptapeptide does show ions especially in the high mass region and several fragment ions are also observed, much more structural information is derived from CAD of the $(M+H)^+$ ion. Not only are the intensities increased of the metastable ion spectrum following CAD, but more peaks are also present in the CAD spectrum. Furthermore, one significant feature is that the metastable ion mass spectrum contains only those peaks which are derived directly from the selected precursor ion. This phenomenon of a direct relationship is used to great analytical advantage (Chapter 7). The drawback of the electric sector scan method is that the limited effective resolution caused by the occurrence of peaks with various widths and of clusters of peaks make an unequivocal mass assignment difficult. A significant advantage to this technique is effective elimination of interfering peaks from impurities.

Enzymolysis coupled with FAB MS is utilized for peptide sequencing. A quadrupole mass spectrometer is outfitted with a FAB source (137) and the performance of a commercial FAB source on a combination Finnigan 3300-INCOS data system mass spectrometer is described. Operational and performance details are described with reference to modifications to the signal output circuits, overall sensitivity, mass calibration of the data system, and contribution of the glycerol matrix to general background and specific mass peaks which contaminate the sample spectrum. Sample is bombarded while residing on a teflon-coated steel probe which is machined at the appropriate angle-of-incidence with respect to the projectile. The angular spread of the fast atom beam is measured by exposing a photographic film which is appropriately mounted on the end of a probe located 28 mm from the exit anode of the FAB gun. The diameter of the exposed area is approximately 5 mm and therefore, the dispersion angle of the fast atom beam is calculated to be about 5°. The preamplifier of the signal output circuit is modified to prevent damage to the amplifier by high voltage arcs which occassionally occur. The high background generally produced by FAB is due mainly to reflected neutrals striking the electron multiplier, noise caused by fluctuations of the fast atom beam, and ions derived from both the glycerol matrix and other components in the sample. In the special case of the quadrupole mass spectrometer, the effect of scattered neutrals in producing the high background is great, because of the "line-of-sight" arrangement of the source exit aperture and multiplier. This background problem can be minimized by feeding a variable compensation voltage to this circuit.

A FAB quadrupole MS instrument is utilized to follow catalytic enzymatic reactions in a real-time mode inside the instrument (138). Enzyme catalysis is monitored by following the decrease in the abundance of the $(M+H)^+$ ion of the substrate and the concomitant increase of the $(M+H)^+$ due to the product. Under specific experimental conditions, enzymes retain their activity, even after limited exposure to bombardment by a fast atom beam. For example, trypsin is spotted onto a probe tip, inserted into the ion source, and exposed to the xenon atom beam for periods of up to eight minutes. After exposure, the trypsin is removed and photometrically assayed. Approximately 50% activity of the enzyme remains after an exposure time of three minutes. Trypsin-catalyzed release of para-toluenesulfonylarginine from its methyl ester (TAME) is a normal assay substrate for trypsin. This substrate is monitored with FAB MS. The $(M+H)^+$ ion of TAME (343) decreases, while that of the corresponding free acid (329) increases. Similarly, the enzyme DAP I was studied with glycylphenylalanine beta-naphthylamide as a substrate. Again, the enzyme activity remains during analysis, and reaction goes virtually to completion within 35 minutes. In a third experiment, yeast carboxypeptidase Y is used to hydrolyze [4]Val-Angiotensin III. After approximately 13 minutes of hydrolysis, the most intense ion corresponds to the species which is produced by formation of the peptide which lost phenylalanine from the

C-terminus. Lastly, it was found that yeast (<u>Accharomyces</u> <u>cerevisiae</u>) shows no loss in viability after FAB exposure when compared to controls. Data indicate that, after exposure to FAB conditions for 0.5 min., 78% of the cells remain viable, while a decline to 50% cell viability occurs after 1.33 min. of exposure to the beam. No cells show viability after 8 min. continuous exposure to FAB. Substantial loss of viability results just from the high vacuum exposure alone, although not as great as that loss which is obtained from exposure to the xenon atom beam.

Oligopeptides are subjected to partial enzyme hydrolysis to produce mixtures of subpeptides (139). This mixture can be sampled directly because there is no interference from the enzyme. Qualitative results of two hexapeptidases (carboxypeptidase Y and microsomal leucine-aminopeptidase) are discussed.

With the advent of FAB MS, biologically important compounds having molecular weights up to 6,000 and 9,000 mass units can be analyzed. It is now possible to observe either the mass-defect or mass-excess effects manifested because of the presence of a corresponding larger number of atoms in that molecule. For example, the polypeptide glucagon has a molecular composition of C_{153} H_{224} N_{42} O_{50} S (123). In the mass spectra of compounds having such a high molecular weight, one must clearly distinguish the nomenclature among several pertinent experimental parameters which are related to mass: nominal mass (no fractional mass), monoisotopic mass (most abundant isotopes plus mass defect), that mass having the highest corresponding intensity, and average mass. Accuracy in the usage of this terminology is important because, for example in the case of glucagon, the monoisotopic mass differs by over two mass units from the mass of highest corresponding intensity.

FAB MS has been utilized to elucidate various molecular forms of the opioid peptide dynorphin in bovine adrenal medullary tissue (140). One immunoreactive fraction is identified as $dynorphin_{1-12}$ (MW=1,475) and fraction 2 is a mixture of $dynorphin_{1-11}$ (MW=1,362), $dynorphin_{1-13}$ (MW=1,603), and two modified forms of $dynorphin_{1-13}$ (MW=1,617 and 1631). Most probably, a single or double replacement of alanine for glycine has occurred in the two latter cases. Other possibilities include monomethylation of tyrosine or a replacement of an amino acid between residues 1-6 to give the indicated molecular weight.

6.9 COLLISION ACTIVATED DISSOCIATION MASS SPECTROMETRY

Events in our physical world occur within a given timeframe. With MS, an ion is formed by either removing or adding one electron from/to an even-electron molecule within that amount of time (fsec) that it takes a bond to oscillate. The internal energy of the ionized and generally excited molecule is redistributed within several fsec oscillations in an attempt to stabilize the resultant charge (141). Insufficient redistribution of the internal energy of the ionized molecule may cause fragmentation of a bond within the microseconds that it takes an accelerated ion to traverse a mass spectrometer's electric and magnetic sectors and to strike the first dynode of an electron multiplier detector.

6.9.1 Unimolecular metastable decompositions

A unimolecular metastable decomposition occurs whenever an ionized molecule fragments during the transit time between acceleration in the ion source and before the multiplier detector (142, 143). The ionized species may not be sufficiently stable to remain intact for the microseconds that it takes for acceleration, mass analysis, and detection of an ion. These metastable ion decompositions are observed in a magnetic scan as low intensity, diffuse peaks which are centered at the following (usually fractional) mass:

$$m^* = (m_2)^2 \ / \ m_1 \tag{1}$$

where m^* is the observed metastable peak, m_2 the mass of the product ion, and m_1 the mass of the precursor ion for the reaction:

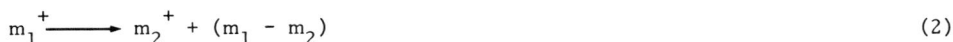

$$m_1{}^+ \longrightarrow m_2{}^+ + (m_1 - m_2) \tag{2}$$

The m^* peak has a Gaussian shape because the unimolecular metastable decompositions have a probability of occurring anywhere along the entire length of the first or second field-free regions (FFR) of a double-focusing mass spectrometer. Figure 6.14 is a box representation of the geometries of two types of double-focusing mass spectrometers. In configuration 1, a forward geometry instrument, the electric sector (E) precedes the magnetic sector (B) while in configuration 2, which represents a reverse geometry instrument, the two fields are reversed. In either configuration, two FFR's are present as indicated, and metastable decompositions may occur in either one or both of these FFR's. To a certain extent, the probability that dissociations will occur in each one of those FFR's is a function of their lengths. Interpretation of these metastable ions facilitates an understanding

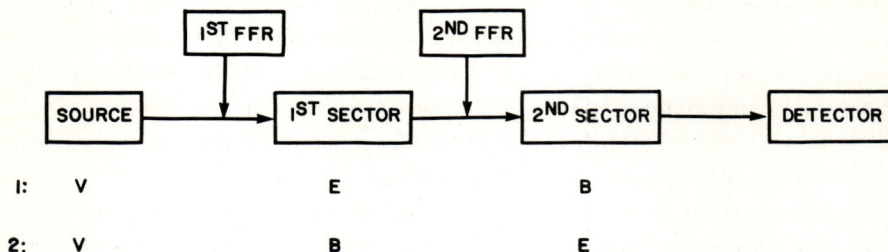

Fig. 6.14. Box diagrams representing the schematics of double-focusing mass spectrometers. Configuration 1 represents a forward geometry (E,B) instrument and configuration 2, a reverse geometry (B,E) instrument.

of their genesis, fragmentation, and structural elucidation because they firmly establish the product/precursor relationship. Establishment of that relationship plays a crucial role and can be used to great analytical advantage.

6.9.2 <u>Collision activated dissociation processes</u>

In CAD spectroscopy, ions having a high translational energy (1-100 keV) collide inelastically with neutral atoms. Part of their translational energy is converted into excitation energy, leading to fragmentation of the excited ions. The major process at energies above 1 keV and scattering angles close to zero is thought to be a vertical electronic excitation of the impacting ion which occurs essentially without momentum transfer to the neutral neutral gas. The excitation energy is rapidly converted to vibrational energy and is distributed statistically over the whole ion, ultimately leading to fragmentation. This energy transfer mechanism is reminiscent of the process occurring in the EI of the neutrals and can be described in terms of the quasi-equilibrium theory (QET). In both cases, broad distributions of internal energies (0-10 eV) are produced, and the spectra obtained are also similar. The amount of translational energy which is converted to internal energy is so large compared to the initial energy of the ions that the CAD spectrum is believed to be unaffected by the latter, although this point has been subjected to debate (144). During subsequent unimolecular dissociation of the excited ion, part of the excess energy is released as translational energy which results in broadening of signals in a CAD spectrum which is obtained by scanning the electrostatic sector.

In a review of the gas phase chemistry of CAD, it is noted that this area of research is as old as mass spectrometry (145). The first mass spectroscopist, Thompson, in 1913, observed the decomposition of collisionally activated $H_2^{+\cdot}$ ions and in 1919, Aston correctly explained the phenomenon. The value of the CAD technique for structural elucidation of organic ions in the gas phase is recognized and the potential of this method is systematically explored for analytcal work (27), in particular for direct analysis of mixtures of organic compounds (134).

The available MS instrumentation will generate ions of either high (keV) or low (eV) translational energy. The following section discusses data for only high energy ions (146). The excitation mechanism for high energy ions derives from the following parameters. The time of collision for a 10 keV ion of mass 100, assuming an interaction distance of 10 angstroms, is approximately 10 fsec. This collision time is comparable to that time for the fastest dissociation which can occur after one vibrational period (10 fsec). Except for the very fastest fragmentations which may occur, these temporal considerations demonstrate that the ion and target are well-separated before dissociation occurs, and that the whole CAD process can be considered to involve two distinct physical steps – the collisional excitation (eq. 3), a process which is always endothermic, and the subsequent unimolecular dissociation (eq. 4), which is exothermic. For a polyatomic ion, we may schematically represent these two steps, respectively, as follows:

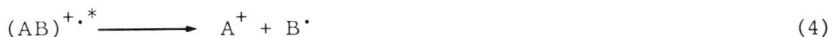

$$(AB)^{+\cdot} + M \longrightarrow (AB)^{+\cdot *} + M \qquad (3)$$

$$(AB)^{+\cdot *} \longrightarrow A^+ + B^\cdot \qquad (4)$$

where $(AB)^{+\cdot}$ is an ionized molecule, AB^{+*} is the corresponding collisionally activated ionized molecule, M is a target molecule, A^+ is a fragment ion, and $B^\cdot$ is a fragment radical.

The inelastic collision of a high energy ion with a neutral target usually leads to an initial electronic excitation. As an ion in its electronic ground state approaches a target molecule M, a strongly repulsive collision complex results. If the ion and the target approach each other infinitely slowly, then the electron density of the system has sufficient time in which to adjust adiabatically in an effort to maintain the lowest total energy which is possible for the system. When the ion and target separate, the ion will be left in its electronic ground state. On the other hand, if the ion-target distance decreases sufficiently rapidly (for example, at that speed which occurs in a high energy collision), then the electrons do not have sufficient time to adjust adiabatically in the molecule's excited state. As a result of the very short interaction time in the collision complex, nuclear motions can be ignored and electronic excitations are considered independently, depending on the relative position of the potential curves. The total cross-section for electronic excitation and the energy transferred during the collision will increase with the ion's translational energy until all statistically significant crossings occur.

The energy transfer and the collisional energy deposition function may be stated as follows:

$$\Delta E_{max} = h/a(2eV/m)^{\frac{1}{2}} \tag{5}$$

where ΔE_{max} is the most probable energy transferred during an inelastic high energy collision, h is Planck's constant, a is interaction distance, e is electronic charge, V is acceleration voltage, m is mass, and v is velocity.

Therefore, for a given interaction distance a, ΔE_{max} is directionally proportional to the square root of the ion's translational energy and inversely proportional to the square root of mass. For a typical interaction distance (usually assumed to be in the range 40-70 nm), the most probable energy of an ion of mass 100 is approximately one eV at four keV. The strong decrease of the energy transfer with the increase in the mass of the colliding ions is critical when one studies molecules of higher molecular weights as, for example, biologically important molecules.

In general, the decomposition of collisionally activated polyatomic organic ions is described within the framework of the QET. Electronically excited ions undergoing rapid radiationless processes revert to the electronic ground state, while the excitation energy is redistributed over all vibrational degrees of freedom. The rate of dissociation is small relative to the rate of energy redistribution, and this fact explains why an isotropic distribution of fragments is observed. In support of this assumption, the same types of fragments often occur and very similar abundances are observed following CAD or EI excitation of a molecule or ion which is also due to similarities in the energy distribution and on ion lifetimes.

The major types of reactions which can occur when a singly-charged positive ion in a keV range collides with a neutral target include the following:

(i)   collision excitation followed by dissociation:

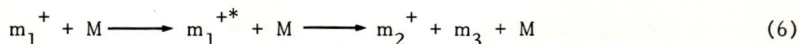

$$m_1^+ + M \longrightarrow m_1^{+*} + M \longrightarrow m_2^+ + m_3 + M \tag{6}$$

(ii) scattering:

$$m_1^+ + M \longrightarrow m^+ \text{ (scattered)} + M \tag{7}$$

(iii) charge exchange (neutralization):

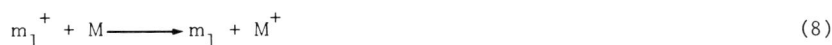

$$m_1^+ + M \longrightarrow m_1 + M^+ \tag{8}$$

(iv) charge stripping:

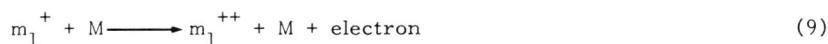

$$m_1^+ + M \longrightarrow m_1^{++} + M + \text{electron} \tag{9}$$

(v) charge inversion:

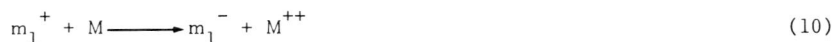

$$m_1^+ + M \longrightarrow m_1^- + M^{++} \tag{10}$$

Reactions 6-10 are in competition with each other, and all will lead to an attenuation of the primary ion beam. The cross-sections for the latter two processes, charge stripping and charge inversion, are smaller relative to the first three collision processes.

It is not the intent of this book nor of this chapter section to delve into the greatest possible depths of CAD MS. However, the following brief discussion provides an appreciation of the wide range of analytical applicability which is available with these CAD processes. The first process, collision excitation followed by dissociation, will be discussed and exemplified in greater analytical detail in Chapter 7.

6.9.3 Scattering

Scattering will lead to a decrease in the intensity of the primary ion beam if the ions are scattered beyond the acceptance angle of the mass spectrometer. Elastic scattering is the result of a Coulombic repulsion between the shielded nuclei of both ion and target. Theoretical considerations predict that the cross-section for elastic scattering decreases with an increase in the ion velocity and a decrease in the size of the target atom. Therefore, for example, instruments operating at higher ion acceleration voltages (8-10 keV) are preferable, and helium is a better target molecule for CAD processes involving keV collisions. On the other hand, for collisions in the eV range, large target masses are preferable. Furthermore, the amount of scattering increases with collision gas pressure.

6.9.4   Charge exchange neutralization

Neutralization of the primary ion beam by charge exchange is the second most important ion removal process, and the cross-section for this collision process may be of the same order as, or even higher, than that for the CAD process itself. Thus, it is possible that, after collision, up to 40% of the initial ion current is due to fast neutrals.   Neutralization of the primary ion beam is minimized, again, if helium is used as target gas.

The target gas pressure has a dramatic influence on the overall CAD efficiency.   Depending on the system studied and the instrument employed, the CAD efficiency passes through a maximum at a target gas pressure which corresponds to a transmittance of 20-60% of the original primary ion beam. This transmission value corresponds to those experimental conditions wherein an ion in the primary ion beam undergoes multiple collisions (on the average, 1.5-2.5 collisions).

CAD efficiency is affected by the fact that the transmission of the CAD fragments arising from the collision region and being transmitted to the detector is lower than that transmission of the primary ion beam because of the greater beam divergence of the fragment ions.   This greater beam divergence is the result of the kinetic energy which is released upon dissociation, but it is also due to the increased scattering of the fragments of smaller mass.   Proper instrumental design may enhance this transmission efficiency, where enhancement of this important parameter can be achieved by confining the collision gas to a very small region near the focal point of the mass spectrometer, and/or by employing mass analyzers with a large acceptance angle for CAD fragments.   Also, increasing the translational energy of the ion will decrease losses due to the angular distribution and increase the mass resolution.

6.9.5   Internal energy effects

Determining ion structures using CAD assumes that the relative abundance of fragments which are generated in CAD processes reflects exclusively the structure of the precursor ion, and is not influenced by its internal energy prior to collision (however, see 145).

For gas phase ion chemistry studies including elucidation of ion structures, decomposition mechanisms, and analytical applications, the fragment's relative abundance is the most important information that can be obtained from a CAD experiment.

6.9.6   Related methods: angle-resolved mass spectrometry

A wide range of energies is deposited into a polyatomic ion during a high energy collision, where the most probable energy transfer under typical experimental conditions is of the order of one electron volt. An approximate selection of the internal energy can be achieved by studying the CAD of a given species under different and well-defined scattering angles. This process is known as "angle-resolved MS" (147). Angular resolution of CAD products can be obtained readily with a conventional double-focusing mass spectrometer of either a normal or reverse Nier-Johnson geometry by utilizing a system of slits or apertures that can define the scattering angle ($0°$ - $2°$) and angular resolution (typically $0.1°$). As generally observed with the interaction between the two important experimental parameters resolution and sensitivity, increased angular resolution is achieved at the expense of sensitivity. The angular distribution of CAD products is due not only to scattering, but may also be due to the release of translational energy during fragmentation.

6.9.7 <u>Analytical applications of collision activated dissociation MS</u>

CAD processes are applied to the elucidation of organic ion structures (148a). The structure of an ion is characterized in this process by matching the CAD mass spectrum with the corresponding spectrum of a known reference compound. Increasing the pressure in the FFR of a mass spectrometer causes collisions to occur between the accelerated ion and neutral atoms, where some of the ion's kinetic energy is transferred into the ion's internal energy. In contrast to metastable ions, the internal energy of the precursor ion causes very little change in the CAD mass spectrum. In this study, a reverse geometry instrument is used, the magnetic field is adjusted to transmit the selected precursor ion, and the unimolecular metastable decomposition mass spectrum is obtained by decreasing the electrostatic analyzer potential (E) from the normal value down to zero, while collecting the transmitted metastable ion current. The pressure in the FFR between the two sectors is then increased with helium until the precursor ion intensity decreases to 10% of the original value. The CAD mass spectrum is determined in a corresponding second electrostatic analyzer scan. Several ions ($C_2H_5O^+$, $C_3H_8^+$, and $C_{13}H_9^+$) are studied in this fashion.

CAD MS unambiguously differentiates between Leu and Ile amino acid residues in peptides by using an immonium ion which is found in the mass spectra of all oligopeptides which contain Leu/Ile, regardless of the relative position of these two amino acid residues in the peptide (148b). Amino acid sequence-determining ions in the EI mass spectra of these peptides are created by cleavages on either side of the carbonyl group of the peptide bond, with charge retention by the N-terminal-containing portion. The CAD mass spectra have peaks of higher abundance, more amino acid sequence-determining peaks, and fewer rearrangement peaks. Furthermore, while collision broadening reduces the effective resolution in CAD processes, a higher ion accelerating voltage is advantageous for CAD spectra to overcome that broadening effect (see above).

In another study, both unimolecular and CAD processes are employed in a MIKES study utilizing both EI and CI methods (149). A series of simple dialkylketones, products of a sterically-controlled organic reaction, and isotopically-enriched species was studied. For example, the McLafferty rearrangment process was not enhanced with ketones during CAD processes. The absolute limit of detection sensitivity obtained for the $(M+H)^+$ ion of p-nitrophenol is quite high, ten pg, and a sample flux of three pg sec^{-1} is noted.

A variety of atomic collision gases (air, helium, argon) and molecules (benzene) is used in a study of 2-butanone, where it was noted that the effect of the different gases on high pressure spectra was small. In an analysis of mixtures of deuterated (d_o, d_1, d_2) diisopropyl ketones by EI, one mixture was chosen for analysis because the compounds in that mixture are isomeric and even the accurate mass selection feature which is possible in a reverse geometry instrument do not separate the components. Nonetheless, the MIKES spectra of the three compounds are readily distinguished. These workers note that a large increase in sensitivity for CAD processes would be obtained by confining those collisions to a very small cross-section chamber which is located at the focal point of the energy analyzer, and that quantification of known compounds could be improved by using isotopically-labeled compounds as internal standards.

Three types of analytical information may be obtained from a CAD mass spectrum:

(i) the relative intensities of the secondary ions;

(ii) the translational energy released upon dissociation; and

(iii) the excitation energy transferred.

While the nature of the collisions has no influence on the intensity ratios, collisions do influence the yield of CAD fragments. As the mass of the atoms used for the collision gas decreases, the cross-section and therefore the yield of the product ion increases. Because of this sensitivity phenomenon, helium and hydrogen are particularly suitable collision gases. While nothing has been reported yet, it is possible that the chemical reactivity of the collision gas might also play a role. While CAD processes can be observed in any mass spectrometer with a magnetic field, a double-focusing, reverse Nier-Johnson geometry is particularly well-suited for this selected purpose because the ion source is followed first by the magnetic field, which performs mass analysis, and then the electric sector, which is scanned to produce the MIKES spectrum. In a forward geometry instrument, linked-field scans (discussed below) are utilized are used to obtain comparable data.

6.9.8 Dimensions of the collision cell

Most conventional collision cells for high energy MS/MS are constructed of a volume of a few cm in length with slits used for both ion entrance and exit. The disadvantage of this type of collision cell is the fact that the collision gas may effuse out from the two slits and into the ion axis. This phenomenon increases the effective length of the collision cell which causes a loss of the required single point of focus. Ideally, an infinitely thin collision region at the object focus point of the mass spectrometer is needed and, towards that end, an efficient collision region using a hypodermic needle has been devised (150). The maximum CAD efficiency for eight keV ions in that system is almost double that of the standard collision cell.

6.9.9 Field desorption - collision activated dissociations

The CAD mass spectra of field-ionized molecules including the acetylated, permethylated tripeptide Leu-Gly-Gly were studied (151). Mainly vertical electronic excitation occurs with the kinetic energy utilized in these mass spectrometers (above one keV) which use a small target atom such as helium. In addition to the tripeptide, several aromatic compounds and two mixtures (estrone and progesterone; phenacetine and methylsalicylate) were studied. These authors state that direct identification of the compounds in the mixture may be possible without any prior chromatographic separation. This concept of greatly increased molecular specificity is quite unique and represents a significant analytical feature which is more fully elaborated in Chapter 7. Major disadvantages noted for CAD MS are the decreased mass resolution due to the kinetic energy released under collision processes and the need in the past for approximately one mg of compound.

Nine dipeptides and two tripeptides were studied utilizing FD in conjunction with CAD processes (152). While the FD mass spectra of the two isobaric and isomeric dipeptides glycylalanine and alanylglycine are similar, the CAD mass spectra of the two protonated isomers differ significantly.

6.9.10 Triple stage quadrupoles

A multianalyzer mass spectrometer constructed from three sets of quadrupoles Q1, Q2, and Q3 was used to study the CAD of the negative anions of several classes of compounds (ketones, esters, nitrophenols, and sugars) (153).

Figure 6.15 contains a box diagram schematic representation of a triple stage quadrupole (TSQ) mass spectrometer configuration. While several computer-controllable operational modes are possible in this combination instrument, Q1 typically operates as a mass selector to transmit a selected ion mass to Q2. The second quadrupole is pressurized and operates as a collision cell. CAD occur at low energy in Q2 to produce a mass spectrum of that ion selected from the mass spectrum produced in Q1. This process is also known as MS/MS. Finally, Q3 obtains the mass spectrum of the CAD products produced in Q2 of that ion selected from Q1. As mentioned in the basic principles section above, a collision gas having a higher atomic weight is more advantageous to use with the lower energy ions found in a quadrupole MS.

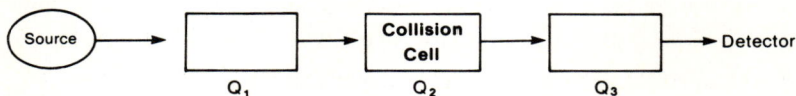

Fig. 6.15. Simplified scheme of a triple stage quadrupole mass spectrometer.

It is significant to note that direct analysis of an industrial sludge containing several priority pollutants is achieved at the 100 part per billon level without any chromatographic extraction. In contradistinction to positive ion MS, the production of $(M-H)^-$ anions proceeds with very little fragmentation, because the energetics of negative ions is small. Negative ion chemistry in the gas phase is complementary to positive ion data. A TSQ mass spectrometer is utilized where the second quadrupole is generally operated as the CAD cell with only RF voltage applied to the rods to transmit all of the ions from Q1. Formation of fragment ions in CAD processes in the TSQ increases with the collision gas in the order: Ar, N_2, Ne, and He. As mentioned above, the interaction time between the ion and collision gas atom or molecule at energies near ten keV is in the fsec range, a time-scale that approximates that time for electronic transitions. Accordingly, it is thought that CAD processes involve an initial vertical electronic excitation which is followed by relaxation and distribution of the excess energy into vibrational modes leading to bond fragmentation. On the other hand, the translational energy of an ion undergoing CAD in a quadrupole is only 5-15 electron volts, and the interaction time between the ion and collision gas atom or molecule is on the psec time scale, which is a time compatible with molecular vibrations. Therefore, the CAD process can occur by direct conversion of translational energy into vibrational energy.

A TSQ mass spectrometer is utilized in the direct amino acid sequence determination of complex mixtures of oligopeptides (154). N-acetylated and N, O-permethylated oligopeptides are analyzed under CI conditions using isobutane as the reagent gas. The first analyzer transmits an $(M+H)^+$ ion relating to a particular oligopeptide. A second analyzer, operating with only low RF potential on the quadrupole rods, acts as an ion-focusing device and transmits all ions of all mass values. In this second analyzer, nitrogen is used to vibrationally excite the $(M+H)^+$ ions and induce their CAD to produce amino acid sequence-determining fragment ions. The third analyzer operates in the normal AC/DC mode to yield the mass spectrum of that product ion selected to be transmitted by the second quadrupole. A series of 26 oligopeptides was studied in this manner in both the normal and CAD modes.

Three limitations to previous work are noted by these authors: low abundance of ions in the high mass region due to the molecular ions of derivatized oligopeptides; the lower efficiency of both the production and collection of CAD ions in a double-focusing mass spectrometer; and the lower mass resolution available for measuring CAD products. The authors state that the TSQ configuration overcomes these three limitations by using CI to generate abundant $(M+H)^+$ ion currents, maximizing the ion dissociation and fragment ion collection efficiency by utilizing low energy collisions inside a quadrupole, and use of a quadrupole instrument to effect unit resolution in the CAD mode, respectively. By utilizing a mixture of d_o- and d_3-N-acetylated peptides, doublets are observed for all N-terminus-containing fragment ions. Amino acid sequence-determining data for glucagon which had been digested with the enzyme elastase and the amino acid sequences for 24 oligopeptides (22 of the 28 peptide bonds) were obtained. A limitation to this methodology is the quadrupole rods which limit analysis to a mass range of up to 1,000 mass units. Lastly, this report contains a scheme to use amino acid sequence-determination data for measurement of neuropeptides at the pmol level utilizing electron capture negative ion mass spectrometry and CAD of the dipeptide carboxylate anions at the 100 pmol (5.5 ng) level. This sensitivity level is comparable to FAB-CAD-B/E methods (Chapter 7).

The combination of SIMS and CAD MS is utilized for the amino acid sequence determination of oligopeptides (155). Mass spectra are obtained on a Finnigan TSQ mass spectrometer with both an ion gun and a saddle-field neutral beam gun. A modified Finnigan LC-MS moving belt interface (see

Chapter 8.6.4) is utilized. A tetrapeptide (five nmol of Met-Arg-Phe-Ala) in a glycerol matrix was treated with oxalic acid. Glycerol minimizes analyte damage while oxalic acid protonates the peptide and increases the glycerol conductivity. Argon is less efficient as a projectile atom compared to xenon and represents a typical phenomenon in FAB MS, where sensitivity is a function of atomic weight of the bombarding atom. While abundant ions are observed with SIMS, few amino acid sequence-determining ions are produced. Amino acid sequence-determining ions are readily produced by CAD processes in the second quadrupole. In a second experiment, both d_0- and d_3-N-acetylated Met-Arg-Phe-Ala are studied in a mixture of four different peptides by analyzing the $(M+H)^+$ ion due to that tetrapeptide alone. Analysis of the CAD products shows that any one mass spectrum of a peptide is free from the mass spectra corresponding to the other three peptides in the mixture. Finally, a heptapeptide, Arg-Val-Tyr-Ile-His-Pro-Phe is sputtered from a glycerol mixture which is deposited on a modified LC-MS moving belt interface. The belt is drawn through the ion source of the mass spectrometer and FAB collisions occur on the belt surface to produce the recognizable FAB mass spectrum of that heptapeptide.

In addition to the low resolution quadrupole multi-analyzer mass spectrometer described recently, MS/MS can also be effected utilizing combinations of high resolution mass spectrometers (85, 156). A double-focusing MS/MS instrument is described with a unique molecular beam collision region. The collision chamber is unique in that the helium beam profile gives 95% of the ion collisions within a distance of less than one micron. The first mass spectrometer in this combination instrument has a resolution of up to 50,000. Furthermore, ions up to a molecular weight of 3,600 mass units may be separated and accelerated to a collision energy of greater than 10 keV.

6.9.11 Interference peaks

A detailed analysis of the genesis of interference peaks which may occur in a MIKES spectrum is presented (157). In addition to the normal peaks which occur, five other origins of these interference peaks are noted: coincidental masses, further fragmentation, doubly-charged ions, charge stripping processes, and further fragmentation of multiply-charged ions. The interferences are most prominent when the MIKE spectrum is obtained from a minor peak of a relatively lower charged ratio. The loss of a steroid sidechain is studied, wherein several examples of each one of these types of interference ions are noted. A list of features is presented to differentiate artifact peaks from true MIKE peaks: (i) most artifact peaks are narrow; (ii) non-integral mass numbers are observed; (iii) utilizing appropriate equations, both precursor and product ions are identified and to confirm that mass assignment, the mass spectrum should contain those two peaks; (iv) CAD processes in the second FFR should enhance true MIKE peaks, while artifact peaks are unaffected; and (v) a slight variation of the magnetic field will destroy true MIKE peaks, while interfering peaks remain.

6.10   SCANNING METHODS TO OBTAIN METASTABLE TRANSITIONS

A metastable transition corresponds to the decomposition of an ion after it leaves the ion source exit slit and before it reaches the collector slit (143). The product ions arising from these in transito decompositions produce metastable peaks in the mass spectrum which are analyzed by a magnetic deflection instrument. Table 6.1 collects the seven methods which are available to the experimenter to observe metastable transitions according to the general equation (2). No one of these seven methods may be an ideal method, and the analyst must consider what type of analytical information is sought from an experiment before choosing from among these methods. Available information from metastable decompositions includes potential energy and ion current surfaces over which the particular decomposition occurs (158, 159), assignment of the mass of either m_1 or m_2, CAD, sensitivity, scan speed, discrimination effects, and interpretation of relative peak intensities.

6.10.1 Basic concepts

In a double-focusing mass spectrometer instrument, three fields may be varied. The schematic representation shown in Fig. 6.14 illustrates these fields as the accelerating voltage (V), the electric sector field (E), and the magnetic sector field (B). These three fields may be scanned either separately, or in several combined or linked-field methods to collect the product ions formed in those metastable transitions in the FFRs. Figure 6.16 is a schematic representation of a Mattauch-Herzog double-focusing mass spectrometer to demonstrate: the accelerating voltage (8kV); the collision chamber filled with collision gas; the CAD of a selected precursor ion m_1 to products m_2^+ and m_3 in the first FFR; and the two fields B and E.

6.10.2 Basic types of scans to observe metastable transitions

A. Scan of one field

The first method listed in Table 6.1 takes advantage of the small, diffuse, Gaussian-shaped peaks which are commonly observed in the normal magnetic field scan mass spectrum, while keeping V and E constant. The information obtained from this type of measurement is the ratio $(m_2)^2/m_1$ which equals the mass of the metastable peak, m^* (see equations 1 and 2).

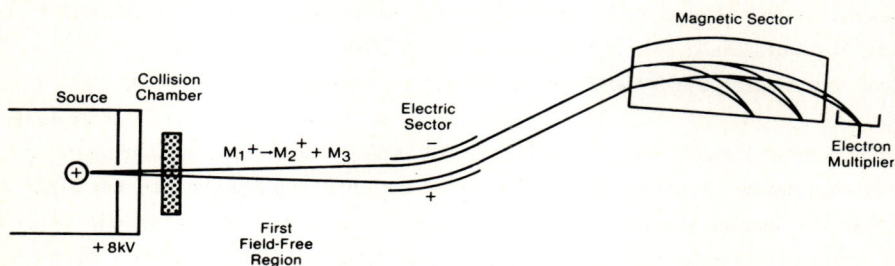

Fig. 6.16. Schematic of a double-focusing E,B mass spectrometer illustrating CAD and decompositions in first FFR.

TABLE 6.1. Mass spectrometric instrumental scanning methods available to observe metastable transitions deriving from the equation $M_1 \longrightarrow M_2^+ + (M_1-M_2)$.

	SCANNED FIELD(S)	FIXED FIELD(S)	INFORMATION OBTAINED
I.	B	V,E	M_2^2/M_1
2.	V	E,B	ALL M_1^+ FROM SELECTED M_2^+
3.	E	V,B	FOR (E, B), ALL M_1^+/M_2^+. FOR (B, E), ALL M_2^+ FROM SELECTED M_1^+
4.	$V^{1/2}/E$	B	ALL M_2^+ FROM SELECTED M_1^+
5.	B/E	V	ALL M_2^+ FROM SELECTED M_1^+
6.	B^2/E	V	ALL M_1^+ FROM SELECTED M_2^+
7.	$(B/E)(1-E)^{1/2}$	V	CONSTANT NEUTRAL FRAGMENT LOSS

In the second method, the accelerating voltage (V) is scanned while keeping the other two fields (E and B) fixed. All of the precursor ions leading to a selected product ion are observed in this type of scan. The spectrum is an energy spectrum of those precursor ions which fragment to yield that selected product ion. Further information can be obtained whenever the fragmentation is accompanied by the release of translational energy. Depending of the width of the beta slit, the peak shape may change and evaluation of the energy release is facilitated. The main advantages of this scan method are the simplicity of the scan and the detection sensitivity which is coupled with the ability to give detailed information on energy release fragmentation. Major disadvantages of the method are that the practical mass scan range is limited and that the source focusing characteristics change during the scan.

In the third method, the electric field E is scanned while keeping the acceleration voltage V and magnetic field B constant. Many of the advantages and disadvantages of the previous method are also observed, especially when the instrument is of a reverse (B,E) geometry. The mass spectrum obtained with this method is known as an ion kinetic energy (IKE) spectrum. However, in a reverse geometry instrument (B,E), the main ion beam is first mass-analyzed by the magnetic sector before the electric sector, decompositions of the precursor ion in the FFR between the two sectors gives rise to various product ions (m_2^+) which are observed by scanning the electric voltage (E), and the resulting spectrum is a MIKE mass spectrum. Two advantages of the MIKES method over the V-scan (method 2) are that the source conditions remain constant throughout the scan and that the range of E_1/E_0 is substantially greater than the range of V_1/V_0 in the previous method. Disadvantages include the fact that, since observations are made later in the ion flight path than with the V-scan method, a smaller number of ions in general will decompose and therefore the sensitivity is lower.

B. Scans where two of the three fields are linked together

The last four methods (4-7) differ from the first three in that two of the three available fields in a double-focusing mass spectrometer are scanned simultaneously to provide a linked-field scan mass spectrum. Each of these linked-field scans produces narrow metastable peaks comparable to the peak width found in the normal mass spectrum. On the other hand, these four linked-field scan methods do not produce any information on the peak shape, and no information on release of translational energy and fragmentation pathway can be deduced.

In the first linked-field scan (method 4), $V^{\frac{1}{2}}/E$ is scanned while keeping the magnetic field strength B constant as a means to collect all of the product ions arising from one selected precursor ion. Two of the disadvantages of this linked-field scan method are limited mass range and the variation of source-tuning conditions during the scan.

In the B/E linked-field scan (method 5), the acceleration voltage V is held constant while the magnetic (B) and electric fields (E) are scanned, maintaining the ratio B/E constant throughout the scan. This linked-field scan mode is used to great analytical advantage, and it will be discussed in

Chapter 7, because all product ions arising from a selected precursor ion are
readily determined. This linked-field scan mode has the advantages that the
source extraction efficiency is constant (V is constant) and that the mass
range scanned is limited only by that range over which the Hall probe is
correlatable to the relationship to the magnetic field strength B. Sensitivity
of this linked-field scan method is higher than the other scan methods because
the ion products arising from decompositions occurring in the first FFR are
collected. However, as in the E-scan (method 3) with a reverse geometry
instrument, the response of the electron multiplier decreases as E decreases
because the relative intensity of peaks within a given linked-field scan will
depend on the energy released for each decomposition.

Another linked-field scan, method (6), is the B^2/E method, which also
keeps the accelerating voltage V constant. This method collects all of the
precursor ions which produce one selected product ion m_2^+. The peak widths
determined with this method are not as sharp as they are in the B/E linked-
field scan method but, nonetheless, the critical precursor-product
relationship is readily established and rigidly maintained.

The last linked-field scan (method 7), in Table 6.1 scans the
relationship $(B/E)(1-E)^{\frac{1}{2}}$ while keeping the accelerating voltage V constant,
and is used to selectively record the first FFR metastable peaks. These
transitions represent the loss of one selected neutral fragment. This type of
"constant neutral" metastable decomposition linked-field mass spectrum is
useful in structural interpretation of the mass spectrum of an unknown
compound (142, 160) or in the analysis of a preselected compound type during
a chromatographic analysis.

In some cases, the diagnostic capability of linked-field scans may be
limited by simultaneous production of peaks which can arise from
fragmentation processes with mass values close to that ion which is selected
for monitoring, thus causing interferences (see discussion above). Alternating
current modulation of either the accelerating voltage V or the electric sector
voltage E, with appropriate phase-shift detection methods, can identify the
artifacts and, in suitable cases, reduce to nearly zero the intensity of those
interfering ions (161).

6.10.3 "Younger" versus "older" ion transitions

An elegant set of experiments was performed in an effort to elucidate
decompositions in the FFR region of a double-focusing mass spectrometer as
opposed to decompositions in the second FFR (162).

The genesis of amino acid sequence-determining ions in tripeptides
which contain an N-terminal trifunctional amino acid was analyzed by direct

analysis of daughter ions/mass analyzed ion kinetic energy (DADI/MIKES) and linked-field scan (B/E) mass spectrometry. CAD in the first and second FFRs of a reversed geometry instrument were compared. The mass spectrum of the tripeptide N-acetyl GluAlaLeu-OMe was investigated in detail. The DADI/MIKE mass spectra show very few differences when compared to the B/E mass spectra, contrary to initial expectations. CAD enhance the number and intensity of fragment ions. In this study, DADI/MIKE spectra are obtained at a magnetic field which is set to transmit one selected ion to the second FFR, in which CAD processes occur. The voltage on the electric sector is scanned and a normal MIKE mass spectrum is obtained with the normally observed broad peaks. On the other hand, "younger" ion transitions are obtained with a linked-field (B/E) scan where a precursor ion is selected and the first FFR decompositions are studied with and without CAD processes. A B/E scan (keeping the B/E ratio constant) is obtained to detect all the product ions arising from that precursor ion. The peaks are comparatively narrow in the B/E scan mode compared to a MIKE scan. Additionally, it must be remembered that those B/E-scanned product ions may still undergo further metastable decompositions in the second FFR.

6.10.4 <u>Selected analytical applications of linked-field scanning mass</u>
<u>spectrometry</u>

Once an ion [generally, the $(M+H)^+$] is produced by an ionization technique (EI, CI, FI, FD, FAB, ^{252}Californium plasma desorption, DCI, etc.), that ion may be subjected to a linked-field scan analysis of metastable transitions. This combination of ion production and linked-field scanning methodology, especially in the case where the intensity of metastable transitions is increased or even created with CAD processes, is a powerful analytical technique whose range of analytical applications is only now being realized. This section will briefly review selectively some of the limited published analytical data employing linked-field scanning techniques for analysis of selected compound classes and Chapter 7 is devoted entirely to the measurement of endogenous opiod peptides. The range of compounds studied includes peptides, tetrahydrocannabinol, steroids, hydroxyphenylacetic acid, and alkaloids.

(i) <u>Peptides</u>

Amino acid sequence-determining data derived from tripeptides and tetrapeptides are analyzed by DADI-MIKE and B/E linked-field scan mass spectrometry. CAD occurs in the first and second FFR of a double-focusing, Nier-Johnson reverse geometry instrument. It has found that CAD enhances the intensity of amino acid sequence-determining ions (163). $(M+H)^+$ ions of several peptides are produced by FD MS. CAD processes increase the intensity of amino acid sequence-determining ions (164) and a computer system accumulates consecutive linked-field scans.

In a similar study utilizing the instrumental combination of FD MS, CAD, and B/E linked-field scanning, several chemically underivatized peptides were studied (100). The opioid peptide LE and the peptides Phe_4 and Phe_2 were examined.

(ii) Tetrahydrocannabinol

A method for a sensitive analysis of tetrahydrocannabinol in plasma was developed by using combined GC-MS with metastable ion monitoring (165). Tetrahydrocannabinol is converted to the trimethylsilyl derivative and then quantified using the metastable ion at mass 371 formed in the transition $M^+ \longrightarrow (M-CH_3)^+$. A tetradeuterated internal standard is utilized and quantification is performed down to the five pg ml^{-1} plasma level. Because of the utilization of the unique metastable transition, much of the chemical, instrumental, and background noise is avoided and the need for extensive purification is bypassed.

(iii) Hydroxyphenylacetic acid

A very high degree of molecular specificity and detection sensitivity is achieved in the linked-field scan mode because ions such as those due to GC column bleed as well as other contaminants in the sample will not be transmitted by the electric sector (166, 167). Both meta- and para-hydroxyphenylacetic acid are measured in rat brain caudate nuclei. Pentafluoropropionyl derivatives are utilized in conjunction with di-deutero analogs, and a specific metastable transition is monitored. Data obtained indicate that the selected metastable peak monitoring method is equivalent to the high resolution- SIM MS technique (168). A detailed discussion indicates that a large (one mass unit) error is noted in the linked-field scan mode and that it is quite important to judiciously select a metastable transition to minimize this mass error. This mass error may be minimized by selecting a metastable transition which does not lose a significant portion of the mass of the molecule, which in turn relates to the partitioning of the original translational energy observed when two fragments form.

(iv) Steroids

Major fragmentation pathways of eight isomeric androstane-3,17-diols plus their bis-t-butyldimethylsilyl ethers were analyzed by linked-field scanning MS techniques. Four metastable pathways were observed, and each is characterized by a loss of an alkylsilanol moiety from a common precursor (169).

(v) Alkaloids

A crude extract from an ergot fermentation broth was studied by a combination of HPLC, CI MS, and the linked-field scan (B/E) method, both with and without CAD processes. Eleven alkaloids are identified in the mixture of at least 18 unknown molecules (170).

6.11 COMPUTER TECHNIQUES

Computer techniques have been developed to record, in a fast and facile manner, the magnetically-scanned, focused metastable ion mass spectra. A time-mass calibration of the mass spectrometer is the basis for calculating both linked-field scans (36, 160).

6.12 ^{252}Cf PLASMA DESORPTION MASS SPECTROMETRY

Basic research has gone into studies of the ionization process of large molecules where for years, researchers in the field of MS have realized that any method to produce an ion from a compound is appropriate for MS study. Nuclear chemistry has supplied an elegant method for production of ions from large, thermally labile, polar biomolecules such as polynucleotides and polypeptides (171, 172). The basic principles of this ionization system are outlined elsewhere and will not be elaborated here. Briefly, a thin nickel foil is subjected to ^{252}Cf fission fragment bombardment. The ^{252}Cf nucleus decays with a half-life of 2.6 years, 3% of the decay arises via spontaneous fission and 97% by emission of alpha particles. The decay is temporally random, and fission fragments are emitted in all directions. Each fission event produces two fragments which travel in opposite directions. A typical pair of fission fragments is ^{142}Ba^{18+} and ^{106}Te^{22+}, which have kinetic energies of ca. 79 and 104 MeV, respectively. Approximately 40 different pairs of fragments can result from spontaneous fission of ^{252}Cf, and the above pair is representative of the masses, highly-charged states, and kinetic energies which are involved (173). One of the two fission fragments strikes a counter which initiates a time-sweep of a time-of-flight mass spectrometer and also serves as a timing marker. The other fission fragment impacts onto the nickel foil where the sample is deposited as a thin film, and ionization of the sample occurs. The mass spectra produced by these fission fragments are summed over a period of time to produce a time-averaged composite mass spectrum. The types of MS data obtained are impressive, in that compounds with very large molecular weights are efficiently ionized. A fully chemically protected polynucleotide and intact insulin were studied by this technique.

Particle-induced desorption mass spectrometry is a term which includes plasma desorption mass spectrometry. Plasma desorption mass spectrometry has been studied with ^{252}Cf nuclear fission products (171, 172) and ^{127}I plasma desorption (174). More extensive details of this desorption process are discussed elsewhere (171-175).

6.13 FOURIER TRANSFORM MS

The field of Fourier transform (FT) MS has experienced some recent, exciting results and holds significant promise as an analytical technique (176, 177). Resolution values of one part in 1,500,000 (mass 166, 4.7 Tesla) and a mass-measuring accuracy of 0.4 ppm (mass 235) have been attained. Higher mass and greater sensitivity are two other important operational specifications that are constantly being improved in commercially available FT-MS instruments.

6.14 SUMMARY

Each one of the MS amino acid sequencing techniques mentioned above has its own inherent experimental advantages and disadvantages. It is the author's opinion and experience that, when working with an endogenous peptide, it is advantageous to avoid or minimize the number of experimental manipulations such as transferring samples and chemical derivatizations. For example, preference is given to ionization techniques such as FD and FAB MS, because no chemical derivatives are required. Linked-field scanning (with or without CAD) may provide amino acid sequence-determining ions which can be used for structual elucidation of an unknown peptide, or for quantification of a known peptide.

If sufficient material is available, and if appropriate MS instrumentation and computer hardware and software are at hand, then the other experimental techniques such as DAP enzymolysis or random acid hydrolysis are useful for amino acid sequence determination. These two methods are not as applicable for quantification purposes. In principle, more sample would be lost in these two methods which require enzymolysis and/or chemical reactions. The permethylation reaction appears to yield the most side-products. However, because those side-products have been studied, they may also be analytically useful. LC-MS studies of permethylated peptides indicate variability of methylation as well as possible C-methylation products. The method that will be chosen for peptide analysis depends on two critical experimental considerations: first and foremost, the amount of sample and second, whether the objective of the experiment is qualitative or quantitative analysis.

Other ionization methods (FD, FAB, ^{252}Cf, FT-MS) are described and their respective advantages and disadvantages were analyzed. Once an ion is made and accelerated, other instrumental manipulations such as CAD, IKE, MIKE, DADI, and linked-field scans are possible to facilitate ion structure determinations.

REFERENCES

1   K. Biemann, Mass Spectrometry: Organic Chemical Applications, McGraw-Hill, New York, 1962, 370 pp.
2   J.H. Beynon, Mass Spectrometry and its Applications to Organic Chemistry, Elsevier, Amsterdam, 1960, 640 pp.
3   F.W. McLafferty, Interpretation of Mass Spectra, University Science Books, Mill Valley, CA, 1980, 303 pp.
4   F.W. McLafferty and R. Venkataraghavan, Mass Spectral Correlations, American Chemical Society, Washington, D.C., 1982, 124 pp.
5   H. Budzikiewicz, C. Djerassi and D.H. Williams, Mass Spectrometry of Organic Compounds, Holden-Day, San Francisco, 1967, 690 pp.
6   J.T. Watson, Introduction to Mass Spectrometry: Biomedical, Environmental, and Forensic Applications, Raven Press, New York, 1976, 239 pp.
7   B.S. Middleditch (Editor), Practical Mass Spectrometry - A Contemporary Introduction, Plenum, New York, 1979, 387 pp.
8   F.W. Lampe (Editor) Thirtieth Annual Conference on Mass Spectrometry and Allied Topics, Honolulu, ASMS, E. Lansing, MI (1982) 942 pp.
9   N.R. Daly (Editor), Advances in Mass Spectrometry, Vol 7A, Heyden, London, 1978, 814 pp.
10 A. Quayle (Editor), Advances in Mass Spectrometry, Vols, 8A, 8B, Heyden, London, 1980, 1975 pp.
11 A. Quayle (Editor), Advances in Mass Spectrometry, 9, Heyden, in press.
12 G.R. Waller and O.C. Dermer (Editors), Mass Spectrom. Rev., 1 (1982).
13 A.L. Burlingame, J.O. Whitney and D.H. Russell, Anal. Chem., 56 (1984) 417R-467R.
14 S.I. Goodman and S.P. Markey, Diagnosis of Organic Acidemias by GC-MS, Liss, New York, 1981, 158 pp.
15 E. Costa and B. Holmstedt, Gas Chromatography-Mass Spectrometry in Neurobiology, Raven, New York, 1973, 175 pp.
16 A. Frigerio and N. Castagnoli (Editors), Advances in Mass Spectrometry in Biochemistry and Medicine, Vol. 1, Spectrum, New York, 1976, 586 pp.
17 A. Frigerio (Editor), Advances in Mass Spectrometry in Biochemistry and Medicine, Vol. 2, Spectrum, New York, 1977, 609 pp.
18 A. Frigerio and E.L. Ghisalberti (Editors), Mass Spectrometry in Drug Metabolism, Plenum, 1977, 532 pp.
19 G.R. Waller (Editor), Biochemical Applications of Mass Spectrometry Wiley, New York, 1972, 872 pp.
20 G.R. Waller and O.C. Dermer, Biochemical Applications of Mass Spectrometry: First Supplementary Volume, Wiley, New York, 1980, 1279 pp.
21 C.A. McDowell (Editor), Mass Spectrometry, McGraw-Hill, New York 1963, 639 pp.
22 R.G. Cooks, J.H. Beynon, R.M. Caprioli and G.R. Lester, Metastable Ions, Elsevier, New York (1973) 296 pp.

23   R.G. Cooks (Editor), Collision Spectroscopy, Plenum, New York, 1978, 458 pp.
24   M.L. Gross (Editor), High Performance Mass Spectrometry: Chemical Applications, American Chemical Society, 1978, 358 pp.
25   A.L. Burlingame (Editor), Topics in Organic Mass Spectrometry, Wiley, New York, 1970, 471 pp.
26   H. Morris (Editor), Soft Ionization Biological Mass Spectrometry, Heyden, Philadelphia, 1981, 156 pp.
27   F.W. McLafferty (Editor), Tandem Mass Spectrometry, Wiley, NY, 1983, 506 pp.
28   E.R. Klein and P.D. Klein, Biomed. Mass Spectrom., 5 (1978) 425-432.
29   T.A. Baillie (Editor), Stable Isotopes, University Park Press, London, 1978, 314 pp.
30   B.J. Millard, Quantitative Mass Spectrometry, Heyden, Philadelphia, 1978, 171 pp.
31   D.R. Knapp, Handbook of Analytical Derivatization Reactions, Wiley, New York, 1979, 741 pp.
32   M. Spiteller and G. Spiteller, Massenspektrensammlung von Losungsmitteln, Verunreinigungen, Saulenbeleg materielen und einfachen aliphatischen Verbindungen, Springer, New York, 1973, 62 pp plus 170 spectra.
33   M. Windholz, S. Bundavari, M. N. Fertig, and G. Albers-Schonberg (Editors), Table of Molecular Weights, Merck, Rahway, 1978, 257 pp.
34   J.H. Beynon and A.E. Williams, Mass and Abundance Tables for Use in Mass Spectrometry, Elsevier, New York, 1963, 570 pp.
35   H.D. Beckey, Field Ionization Mass Spectrometry, Pergamon, New York 1971, 344 pp.
36   A. Carrick, Computers and Instrumentation, Heyden, London, 1979, 327 pp.
37   K. Biemann, F. Gapp and J. Seibl, J. Amer. Chem. Soc., 81 (1959) 2274.
38   K. Biemann and W. Vetter, Biochem. Biophys. Res. Commun. 3 (1960) 578-584.
39   J.H. Beynon, Mass Spectrometry and its Applications to Organic Chemistry, Elsevier, Amsterdam, 1960, 640 pp.
40   K. Biemann, P. Bommer and D.M. Desiderio, Tet. Lett., 26 (1964) 1725-1731.
41   D.M. Desiderio and T.E. Mead, Anal. Chem., 40 (1968) 2090-2096.
42   D.M. Desiderio, in G.W.A. Milne (Editor), Mass Spectrometry: Techniques and Applications, Wiley, New York, 1970, pp. 11-42.
43   K. Biemann, in G.R. Waller (Editor), Biomedical Applications of Mass Spectrometry, Wiley, New York, 1972, pp. 405-428.
44   H. Frank and D.M. Desiderio, Anal. Biochem., 90 (1978) 413-419.
45   V.K. Mahajan and D.M. Desiderio, Biochem. Biophys. Res. Commun., 82 (1978) 1104-1110.
46   D.H. Thomas, B.C. Das, S.D. Gero and E. Lederer, Biochem. Biophys. Res. Commun., 32 (1968) 519-525.
47   E. Lederer, Pure Appl. Chem., 17 (1968) 489-517.
48   K.D. Haegele, G. Holzer, W. Parr, S.H. Nakagawa and D.M. Desiderio, Biomed. Mass Spectrom., 1 (1974) 175-189.
49   P.A. Leclercq, L.C. Smith and D.M. Desiderio, Biochem. Biophys. Res. Commun., 45 (1971) 937-944.
50   H. Frank, K.D. Haegele, and D.M. Desiderio, Anal. Chem., 49 (1977) 287-291.
51   H. Nau and K. Biemann, Anal. Biochem., 73 (1976) 154-174.
52   P. White and D.M. Desiderio, Anal. Lett., 4 (1971) 141-149.
53   K. Biemann, C. Cone, B.R. Webster and G.P. Arsenault, J. Amer. Chem. Soc., 88 (1966) 5598-6006.

54 M. Senn, R. Venkataraghavan, and F.W. McLafferty, J. Amer. Chem. Soc., 88 (1966) 5593-5597.
55 K. Biemann, in Waller and Dermer (Editors), Biochemical Applications of Mass Spectrometry, Wiley, New York, 1980, pp. 469-525.
56 P.J. Arpino and F.W. McLafferty, in Nachod and Phillips (Editors), Determination of Organic Structures by Physical Methods, Vol. 6, Academic Press, New York, 1976, pp. 1-89.
57 R. Burgus, T.F. Dunn, D.M. Desiderio, D.N. Ward, W. Vale and R. Guillemin, Nature, 226 (1970) 321-325.
58 D.M. Desiderio, R. Burgus, T.F. Dunn, D.N. Ward, W. Vale and R. Guillemin, Org. Mass Spectrom., 5 (1981) 221-228.
59 P.A. Leclercq and D.M. Desiderio, Anal. Lett., 4 (1971) 305-316.
60 S.D. Putney, N.J. Royal, H. Neuman de Vegvar, W.C. Herlihy, K. Biemann and P. Schimmel, Science, 213 (1981) 1497-1501; B.W. Gibson and K. Biemann, Proc. Natl. Acad. Sci., 81 (1984) 1956-1960.
61 E.J. Corey and M. Chaykovsky, J. Amer. Chem. Soc., 84 (1962) 866-868.
62 A. Dell, H.R. Morris, D.H. Williams and R.P. Ambler, Biomed. Mass Spectrom., 1 (1974) 269-273.
63 R.M. Caprioli, W.E. Seifert and D.E. Sutherland, Biochem. Biophys. Res. Commun., 55 (1973) 65-75.
64 M.A. Young and D.M. Desiderio, Anal. Biochem., 70 (1976) 110-123.
65 A.G. Harrison, Chemical Ionization Mass Spectrometry, CRC Press, Boca Raton, FL, (1983), 156 pp.
66 W.R. Gray, L.H. Wojcik and J.H. Futtrell, Biochem. Biophys. Res. Commun., 41 (1970) 1111-1119.
67 P.A. Leclercq and D.M. Desiderio, Org. Mass Spectrom., 7 (1973) 515-533.
68a H.D. Beckey, in G. Waller (Editor), Biochemical Applications of Mass Spectrometry, Wiley, New York, 1972, pp. 795-815.
68b H.-R. Schulten, in Glick (Editor), Methods of Biochemical Analysis, Vol. 24, Wiley, New York, 1977, pp. 314-448.
69 J.W. Maine, B. Soltmann, J.F. Holland, N.D. Toung, J.N. Gerber and C.C. Sweeley, Anal. Chem., 48 (1976) 427-429.
70 D.M. Desiderio, J.L. Stein, M.D. Cunningham and J.Z. Sabbatini, J. Chromatogr., 195 (1980) 369-377.
71 R.C. Murphy, K.L. Clay and W.R. Matthews, Anal. Chem., 54 (1982) 336-338.
72 G.D. Daves, Accts. Chem. Res., 12 (1979) 359-365.
73 H.U. Winkler and H.D. Beckey, Biochem. Biophys. Res. Commun., 46 (1972) 391-398.
74 S. Asante-Poku, G.W. Wood and D.E. Schmidt, Jr., Biomed. Mass Spectrom., 2 (1975) 121-125.
75 H.U. Winkler, R.J. Beuhler and L. Friedman, Biomed. Mass Spectrom., 3 (1976) 201-206.
76 M. Przybylski, I. Luderwald, E. Kraas, W. Voelter and S.D. Nelson, Z. Naturforsch., 34b (1979) 736-743.
77 B. Calas, J. Mery, J. Parello, J.C. Prome, J. Roussel and D. Patouraux, Biomed. Mass Spectrom., 7 (1980) 288-293.
78 W. Frick, G.D. Daves, Jr., D.F. Barofsky, E. Barofsky, G.H. Fisher, D. Chang and K. Folkers, Biomed. Mass Spectrom., 4 (1977) 152-154.
79 W. Frick, E. Barofsky, G.D. Daves, Jr., D.F. Barofsky, D. Chang, and K. Folkers, J. Amer. Chem. Soc., 100 (1978) 6221-6225.
80 J.F. Holland, B. Soltmann and C.C. Sweeley, Biomed. Mass Spectrom., 3 (1976) 340-345.
81 H.D. Beckey and F.W. Rollgen, Org. Mass Spectrom., 4 (1979) 188-190.

176

82   V. Giessmann and F.W. Rollgen, Int. J. Mass Spectrom. Ion Phys., 38 (1981) 267-279.

83   S.S Wong and F.W. Rollgen, Proc. 29th Field Emission Sympos., in Andersen and Norden (Editors), Almqvist and Wiksell, Stockholm, 1982, pp. 225-230.

84   S.S. Wong, V. Giessmann, M. Karas, and F.W. Rollgen, Int. J. Mass Spectrom. Ion Phys., 56 (1984) 139-150.

85   I.J. Amster, M.A. Baldwin, M.T. Cheng, C.H. Proctor and F.W. McLafferty, J. Amer. Chem. Soc., 105 (1983) 1654-1655.

86   M.G. Darcy, D.E. Rogers and P.J. Derrick, Int. J. Mass Spectrom. Ion Phys., 27 (1978) 335-347.

87   P.G. Cullis, G.M. Neumann, D.E. Rogers, and P.J. Derrick, in A. Quayle (Editor), Adv. in Mass Spectrometry, Vol. 8B, Heyden, London, 1980, pp 1729-1738.

88   H.R. Morris, A. Dell and R.A. McDowell, Biomed. Mass Spectrom., 8 (1981) 463-473.

89   T. Matsuo, H. Matsuda, I. Katakuse, Y. Wada, T. Fujita and A. Hayashi, Biomed. Mass Spectrom., 8 (1981) 25-30.

90   I. Katakuse, T. Matsuo, H. Matsuda, Y. Shimonishi, Y-M Hong and Y. Izumi, Biomed. Mass Spectrom., 9 (1982) 64-68.

91   Y. Shimonishi, Y.-M. Hong, I. Katakuse and S. Hara, Bull. Chem. Soc. Jpn., 54 (1981) 3069-3075.

92   K.L. Olson, K.L. Rinehart, Jr. and J.C. Cook, Jr., Biomed. Mass Spectrom., 4 (1977) 284-290.

93   G.W. Wood and W.F. Sun, Biomed. Mass Spectrom., 7 (1980) 399-400.

94   W.D. Lehmann and H.-R. Schulten, Biomed. Mass Spectrom., 5 (1978) 208-214.

95   P. Vouros, D.M. Desiderio, J. Leferink, J.A. McCloskey, Anal. Chem., 42 (1970) 1275-1277.

96   P. Vouros, D.M. Desiderio, J. Leferink, T.J Odiorne and J.A. McCloskey, Int. J. Mass Spectrom. Ion Phys., 10 (1972) 133-142.

97   A. Shepp, R.E. Whitney and J.I. Master, Photogr. Sci. Eng., 11 (1967) 322.

98   H. Miyazaki, E. Shirai, M. Ishibashi, K. Hosoi, S. Shibata and M. Iwanaga, Biomed. Mass Spectrom., 5 (1978) 559-565.

99   F.S. Tanzer, D.M. Desiderio, and S. Yamada, in D.H. Rich and E. Gross (Editors) Peptides: Synthesis-Structure-Function, Pierce Chemical Co., 1981, pp. 761-764.

100   D.M. Desiderio and J.Z. Sabbatini, Biomed. Mass Spectrom., 8 (1981) 565-568.

101   D.M. Desiderio, S. Yamada, J.Z. Sabbatini and F. Tanzer, Biomed. Mass Spectrom., 8 (1981) 10-12.

102   D.M. Desiderio and M. Kai, Int. J. Mass Spectrom. Ion Phys., 48 (1983) 261-264.

103   A. Benninghoven, Surf. Sci., 35 (1973) 427-457.

104   A. Benninghoven and W. Sichtermann, Org. Mass Spectrom., 12 (1977) 595-597.

105   A. Benninghoven and W.K. Sichtermann, Anal. Chem., 50 (1978) 1180-1184.

106   D.J. Surman and J.C. Vickerman, J.C.S. Chem. Comm. (1981) 324-325.

107   M. Barber, R.S. Bordoli, R.D. Sedgwick and A.N. Tyler, Biomed. Mass Spectrom., 8 (1981) 337-342.

108   M. Barber, R.S. Bordoli, R.D. Sedgwick and A.N. Tyler, Biomed. Mass Spectrom., 9 (1982) 208-214.

109   M. Barber, R.S. Bordoli, R.D. Sedgwick and A.N. Tyler, J.C.S. Chem. Commun., (1981), p. 325.

110 M. Barber, R.S. Bordoli, R.D. Sedgwick, A.N. Tyler, G.V. Garner, D.B. Gordon, L.W. Tetler and R.C Hider, Biomed. Mass Spectrom., 9 (1982) 265-268.

111 M. Barber, R.S. Bordoli, G.J. Elliott, R.D. Sedgwick, A.N. Tyler and B.N. Green, J. Chem. Soc. Chem. Commun., (1982) 936-938; M. Barber, R.S. Bordoli, G.J. Elliott, A.N. Tyler, J.C. Bill, and B.N. Green, Biomed. Mass Spectrom., 11 (1984) 182-186.

112 M. Barber, R.S. Bordoli, G.J. Elliott and N.J. Horoch, Biochem. Biophys. Res. Commun., 110 (1983) 753-757.

113 D. Halliday and R. Resnick, Physics, Wiley, N.Y., 1960, p. 856.

114 J.H. Beynon, D. Cameron and J.F.J. Todd, Int. J. Mass Spectrom. Ion Phys., 42 (1982) 215.

115 J.H. Beynon, D. Cameron and J.F.J. Todd, Anal. Chem., 54 (1982) 679.

116 J. Franks, Int. J. Mass Spectrom. Ion Phys., 46 (1983) 343-346.

117 J. Franks and A.M. Ganhder, Vacuum, 24 (1974) 489-491.

118 S.A. Martin, C.E. Costello and K. Biemann, Anal. Chem., 54 (1982) 2362-2368.

119 J.M. Gillian, P.W. Landis, and J. Occolowitz, Anal. Chem., 55 (1983) 1531-1533.

120 R. Stoll, U. Schade, F.W. Rollgen, V. Giessmann and D.F. Barofsky, Int. J. Mass Spectrom. Ion Phys., 43 (1982) 227-229.

121 S.S. Wong, R. Stoll and F.W. Rollgen, Z. Naturforsch., 37a (1982) 718-719

122 F.H. Field, J. Phys. Chem., 86 (1982) 5115-5123.

123 J. Yergey, D. Heller, G. Hansen, R.J. Cotter and C. Fenselau, Anal. Chem., 55 (1983) 353-356.

124 A. Dell and H.R. Morris, Biochem. Biophys. Res. Commun., 106 (1982) 1456-1462.

125 D.M. Desiderio and I. Katakuse, Biomed. Mass Spectrom., 11 (1984) 55-59.

126 M. Barber, R.S. Bordoli, G.V. Garner, D.B. Gordon, R.D. Sedgwick, L.W. Tetler and A.N. Tyler, Biochem. J., 197 (1981) 401-404.

127 D.H. Williams, C. Bradley, G. Bojesen, S. Santikarn and L.C.E. Taylor, J. Amer. Chem. Soc., 103 (1981) 5700-5704.

128 D.H. Williams, G. Bojesen, A.D. Affret and L.C.E. Taylor, FEBS Lett., 128 (1981) 37-39.

129 D.H. Williams, C.V. Bradley, S. Santikarn and G. Bojesen, Biochem. J., 201 (1982) 105-117.

130 C.V. Bradley, D.H. Williams and M.R. Hanley, Biochem. Biophys. Res. Commun., 104 (1982) 1223-1230.

131 A. Dell, T. Etienne, M. Panico, H.R. Morris, G.P. Vinson, B.J. Whitehouse, M. Barber, R.S. Bordoli, R.D. Sedgwick and A.N. Tyler, Neuropeptides, 2 (1982) 233-240.

132 K.L. Rinehart, L.A. Gaudioso, M.L. Moore, R.C. Pandey, J.C. Cook, M.Barber, R.D. Sedgwick, R.S. Bordoli, A.N. Tyler and B.N. Green, J. Amer. Chem. Soc., 103 (1981) 6517-6520.

133 K.L. Rinehart, Science, 218 (1982) 254-260.

134 K.L. Busch and R.G. Cooks, Science, 218 (1982) 247-254.

135 J.L. Aubagnac and F.M. Devienne, Tetr. Lett., 23 (1982) 5263-5266.

136 W. Heerma, J.P. Kamerling, A.J. Slotboom, G.J.M. van Scharrenburg, B.N. Green and I.A.S. Lewis, Biomed. Mass Spectrom., 10 (1983) 13-16.

137 R.M. Caprioli, C.F. Beckner and L.A. Smith, Biomed. Mass Spectrom., 19 (1983) 94-97.

138 L.A. Smith and R.M. Caprioli, Biomed. Mass Spectrom., 10 (1983) 98-102.

139 R. Self and A. Parente, Biomed. Mass Spectrom., 10 (1983) 78-82.

178

140 S. Lemaire, L. Chouinard and D. Denis, Biochem. Biophys. Res. Commun., 108 (1982) 51-58.

141 K. Levsen and H. Schwarz, Angew. Chem. Int. Ed. (Engl.), 15 (1976) 509-568.

142 K.R. Jennings, in M.L. Gross (Editor), High Performance Mass Spectrometry: Chemical Applications, Amer. Chem. Soc., Washington, D.C., 1978, pp. 3-17.

143 K.R. Jennings and R.S. Mason, in F.W. McLafferty (Editor) Tandem Mass Spectrometry, Wiley, N.Y., 1983, p. 197.

144 C. Dass, Ph.D. Thesis, Univ. Nebraska - Lincoln, 1984.

145 K. Levsen and H. Schwarz, Mass Spectrom. Rev., 2 (1983) 77-148.

146 P.J. Todd and F.W. McLafferty, in F.W. McLafferty (Editor), Tandem Mass Spectrometry, Wiley, NY (1983), p. 149.

147 S.A. McLuckey and R.G. Cooks, in F.W. McLafferty (Editor), Tandem Mass Spectrometry, Wiley, NY (1983), p.303

148a F.W. McLafferty, R. Kornfeld, W.F. Haddon, K. Levsen, I. Sakai, P.F. Bente III, S.-C. Tsai, and H.D.R. Schuddemage, J. Amer. Chem. Soc., 95 (1973) 3886-3892.

148b K. Levsen, H.-K. Wipf and F.W. McLafferty, Org. Mass Spectrom., 8 (1974) 117-128.

149 T.L. Kruger, J.F. Litton, R.W. Kondrat and R.G. Cooks, Anal. Chem., 48 (1976) 2113-2119.

150 G.L. Glish and P.J. Todd, Anal. Chem., 54 (1982) 842-843.

151 K. Levsen and H.D. Beckey, Org. Mass Spectrom., 9 (1974) 570-581.

152 R. Weber and K. Levsen, Biomed. Mass Spectrom., 7 (1980) 314-316.

153 D.F. Hunt, J. Shabanowitz and A.B. Giordani, Anal. Chem., 52 (1980) 386-390.

154 D.F. Hunt, A.M. Buko, J.M. Ballard, J. Shabanowitz and A.B. Giordani, Biomed. Mass Spectrom., 8 (1981) 397-408.

155 D.F. Hunt, W.M. Bone, J. Shabanowitz, J. Rhodes and J.M. Ballard, Anal. Chem., 53 (1981) 1704-1706.

156 F.W. McLafferty, Acc. Chem. Res., 13 (1980) 33-39.

157 T. Ast, M.H. Bozorgzadeh, J.L. Wiebers, J.H. Beynon and A.G. Brenton, Org. Mass Spectrom., 14 (1979) 313-318.

158 M.J. Lacey and C.G. MacDonald, Biomed. Mass Spectrom., 12 (1977) 587-594.

159 M.J. Lacey and C. G MacDonald, in F.W. McLafferty (Editor), Tandem Mass Spectrometry, Wiley, NY, 1983, p. 321.

160 W.F. Haddon, Org. Mass Spectrom., 15 (1980) 539-543.

161 B. Shushan and R.K. Boyd, Int. J. Mass Spectrom. Ion Phys. 34 (1980) 37-62.

162 R. Steinauer and U.P. Schlunegger, Biomed. Mass Spectrom., 9 (1982) 153-157.

163 R. Steinauer, H. Walther and U.P. Schlunegger, Helv. Chim. Acta, 63 (1980) 610-617.

164 T.Matsuo, H. Matsuda, I. Katakuse, Y. Shimonishi, Y. Marvyama, T. Higuchi and E. Kubota, Anal. Chem., 53 (1981) 416-421.

165 D.J. Harvey, J.T.A. Leuschner, and W.D.M. Paton, J. Chromatogr., 202 (1980) 83-92.

166 D.A. Durden, J. Neurosci. Meth., 7 (1983) 61-66.

167 D.A. Durden, Anal. Chem., 54 (1982) 666-670.

168 D.M. Desiderio, in J.C. Giddings, E. Grushka, J. Cazes, and P.R. Brown (Editors), Advances in Chromatography, Vol. 22, Marcel Dekker, New York, 1983, pp. 1-36.

169 S. J.Gaskell, A.W. Pike and D.S. Millington, Biomed. Mass Spectrom., 6 (1979) 78-81.

170 C. Eckers, D.E. Games, D.N.B. Mallen and B.P. Swann, Biomed. Mass Spectrom., 9 (1982) 162-173.

171 R.D. Macfarlane, Acc. Chem. Res., 15 (1982) 268-275.

172 R.D. Macfarlane and T.F. Torgeson, Science, 191 (1976) 920-925.

173 M.L. Vestal, Mass Spectrom. Rev., 2 (1983) 447-480.

174 P.Hakansson, I. Kamensky, B. Sundqvist, J. Fohlman, P. Peterson, C.J. McNeal, and R.D. Macfarlane, J. Amer. Chem. Soc., 104 (1982) 2948-2949.

175 B. Sundqvist, I. Kamensky, P. Hakansson, J. Kiellberg, M. Salehpour, S. Widdiyasekera, J. Fohlman, P.A. Peterson and P. Roepstorff, Biomed. Mass Spectrom., 11 (1984) 242-257.

176 C.L. Wilkins and M.L. Gross, Anal. Chem., 53 (1981) 1662A-1676A.

177 J.F. Holland, C.G. Enke, J. Allison, J.J. Stults, J.D. Pinkston, B. Newcome and J.T. Watson, Anal. Chem., 55 (1983) 1002A-1010A.

Chapter 7

MEASUREMENT OF ENDOGENOUS BIOLOGICAL PEPTIDES WITH MASS
SPECTROMETRY

7.1 INTRODUCTION

Recent instrument modifications and analytical applications in the two
areas of HPLC and MS have been developed to such an advanced degree that
researchers can now routinely take advantage of the distinctly separate, but
mutually enhancing properties of these two significant analytical methodologies.
On one hand, for HPLC, the three crucial experimental features of high
resolution, high speed, and high sensitivity are available for fast and efficient
separation of mixtures of peptides. On the other hand, MS now offers several
significant experimental advances including facile production of $(M+H)^+$ ions
from peptides which have molecular weights in a range up to 10,000 mass units,
direct measurement of underivatized peptides, high sensitivity (pmol, fmol), and
the most critical feature of maximum molecular specificity.

7.2 NEUROPEPTIDE MEASUREMENTS

Because of the developments in these two instrumental methods, researchers
can now apply these analytical techniques to the broad field of neurobiology,
and especially to neuropeptides, where an explosive growth of new concepts,
experimental techniques, and information is occurring.

The purpose of this chapter is to illustrate the combination of many of
the analytical capabilities discussed in the previous chapters, while using
measurements of biologically important endogenous peptides to demonstrate this
process. Towards that end, the following topics will be discussed in this
chapter: basic principles of the MS analytical measurement method; synthesis
of stable isotope-incorporated peptide internal standards; measurements of
endogenous neuropeptides; comparison of analytical data derived from MS, RIA,
and RRA methods; different types of MS measurement modes; measurements of
neuropeptides in several biological tissues and fluid extracts; metabolic
profiling of peptides; calibration curve construction; and a comparison of FD
and FAB MS measurement methods. The results from the analytical measurements
are given as ng peptide g^{-1} of wet weight tissue. The other method is to give
data as ng peptide mg^{-1} protein in the tissue. These two methods are directly
correlatable, because brain tissue contains 10% by weight of protein.

7.3 BASIC PRINCIPLES OF ANALYTICAL MEASUREMENT OF PEPTIDES BY MASS SPECTROMETRY

If one considers the abundant $(M+H)^+$ ion current of a peptide which is produced by either FD or FAB processes, and if an appropriate internal standard is available for that selected peptide which is being measured in a biologic extract, then ion currents of the $(M+H)^+$ ions of the peptide of interest and that of the peptide internal standard may be integrated separately. The ratio of those two integrated ion currents, multiplied by the known amount of internal standard originally added to the biologic extract, yields the amount of endogenous peptide in the biologic extract. This unique, direct, and facile analytical measurement method is timely and significant because it provides, for the first time, the unambiguous molecular specificity which is required for measurement of endogenous peptides. MS measurement of peptides is performed routinely at the part per billion or ng peptide g^{-1} tissue level (1-18). While this sensitivity level corresponds to the pmol level for peptides having a molecular weight of 1,000, recent MS instrumental developments indicate that the fmol level of sensitivity will soon be attained while significantly maintaining maximum molecular specificity. The uncompromising guarantee of maximum molecular specificity of the peptide measurement is the hallmark of the MS methodology which is discussed in this chapter. MS has been used to measure many other endogenous and exogenous compounds, but not yet as extensively for peptides.

On the other hand, if we extend our considerations further, it will be realized that the molecular ion, when considered alone in the mass spectrum of a peptide, conveys no amino acid sequence (structural) information. For example, 555 is the nominal mass, $C_{28}H_{37}N_5O_7$ the elemental composition, and 555.2693 the accurate mass of the molecular ion of the opioid pentapeptide LE (TyrGlyGlyPheLeu). Statistical methods (see Chapter 2) demonstrate that 120 (equals 5!) different amino acid sequence permutations (peptides) are possible for any given five amino acids. LE has two glycine amino acid residues, and in this particular case, fewer than 120 different amino acid sequences are possible. Therefore, a significant increase in molecular specificity is demanded for an objective and structurally unambiguous analytical measurement of that endogenous peptide in a biological matrix. That increase in molecular specificity can only be provided at this time by utilizing as the basis for the analytical measurement, a selected amino acid sequence-determining ion in the mass spectrum of a peptide.

7.4 MASS SPECTROMETRIC MEASUREMENT OF ENDOGENOUS PEPTIDES

It may be possible in some cases to utilize CAD dissociations of the FAB-produced $(M+H)^+$ ion of an endogenous neuropeptide of interest to produce an amino acid sequence-determining fragment ion. In other cases, unimolecular dissociations of the $(M+H)^+$ will produce an appropriate ion or ions, while in other cases, no ions will be produced. This process of CAD, followed by linked-field scanning, actually provides the mass spectrum of a selected ion in a mass spectrum. This process is known as mass spectrometry/mass spectrometry (MS/MS, 19) which is comparable to other combination instrumental techniques such as LC-MS and GC-MS. Figure 7.1 schematically represents this combination MS/MS or tandem mass spectrometry. One mass spectrometer, MS-I, serves as a purification step in a fashion parallel to LC, GC, TLC, or any other chromatographic separation step. The second step represents the structure identification step, a function that the MS serves in the combination instruments LC-MS, GC-MS, etc. A second mass spectrometer, MS-II, provides the mass spectrum (with or without CAD) of one selected ion which is produced by MS-I. The same result can be effected in a forward geometry instrument by utilizing a linked-field scan (see Chapter 6). In the case of peptides, a unique amino acid sequence-determining ion is then used for unambiguous quantitative measurement of an endogenous peptide. In the identification mass spectrum in MS-II shown in Fig. 7.1, the relatively narrow peak widths imply that MS-II is a high resolution instrument. In other cases, broader peaks will be observed.

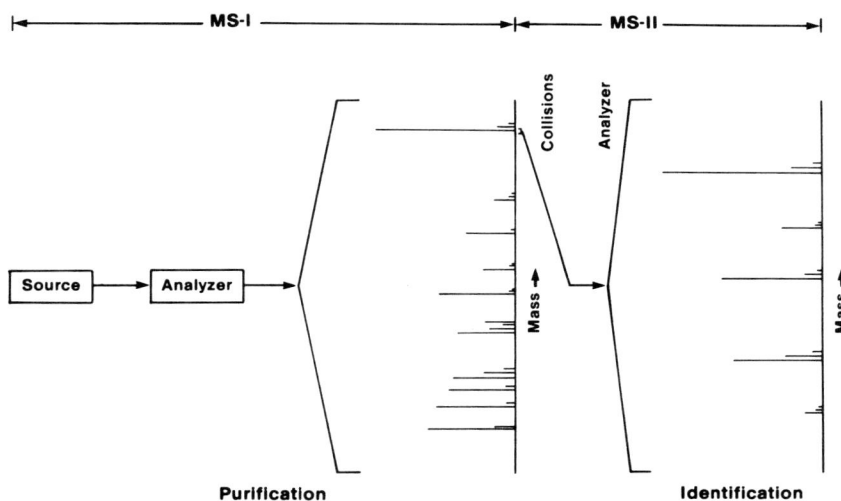

Fig.7.1. Schematic representation of tandem mass spectrometry (MS/MS) (19).

To exemplify the MS/MS process, Figure 7.2 contains the FAB-CAD-B/E mass spectrum of the $(M+H)^+$ ion produced from LE and Figure 7.3 the corresponding spectrum of LE with two ^{18}O atoms located specifically and only in the carboxy terminus (11). Comparing the two mass spectra, it is clear that none of the N-terminus-containing ion masses (221, 278 and 425) shift, whereas all C-terminus-containing ions shift the appropriate four mass units (262, part of 278, 336, 449, 465, 499, and 556). Figure 7.4 contains the fragmentation pattern of LE, and rationalizes the genesis of the ions contained in Figures 7.2 and 7.3. Quite clearly, the C-terminus-containing ions can be used to readily and efficiently differentiate between the two opioid pentapeptides ME and LE (if they were not previously separated by HPLC) by using the C-terminal tripeptide fragment corresponding to -GFL at the two masses 336 (Fig. 7.2) and 340 (Fig. 7.3) for quantification of LE in a biologic extract. That unique amino acid sequence-determining ion is selected as a basis for quantification because it occurs in a noise-free region, there are no neighboring peaks, and it shifts cleanly and completely in the FAB-CAD-B/E mass spectrum (see Fig. 7.3) of the stable isotope-incorporated enkephalin.

7.5 STABLE ISOTOPE-INCORPORATED PEPTIDE INTERNAL STANDARDS

Several types of compounds may be used to serve as appropriate internal standards for MS measurement of endogenous peptides in a biologic extract. First, literally any organic compound can be utilized whenever the detector responses of both the HPLC and MS analytical systems are well-characterized for that compound. However, this type of general internal standard provides neither the highest precision nor accuracy for the important experiment of quantification of an endogenous peptide.

A second class of compounds useful as internal standards is that of structurally homologous compounds. For example, in preliminary peptide measurement studies, ^{2}alanine-leucine enkephalin was utilized as an internal standard (5). However, simply replacing the hydrogen sidechain of a glycine amino acid residue in the LE pentapeptide with the methyl group of alanine significantly increases the hydrophobicity of the pentapeptide molecule and therefore, the RP-HPLC retention time of the homolog increases seven minutes with TEAF buffer and acetonitrile organic modifier (see Chapter 4). While this HPLC separation of peptide and internal standard does not present any major analytical problem, two separate chromatographic fractions must be collected in one reactivial, a procedure that could decrease the accuracy and precision of a measurement.

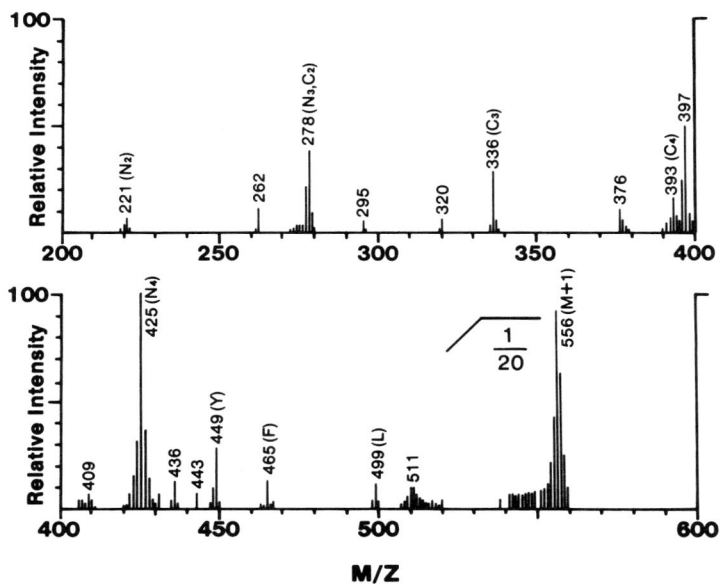

Fig.7.2. FAB-CAD-B/E mass spectrum of LE (8).

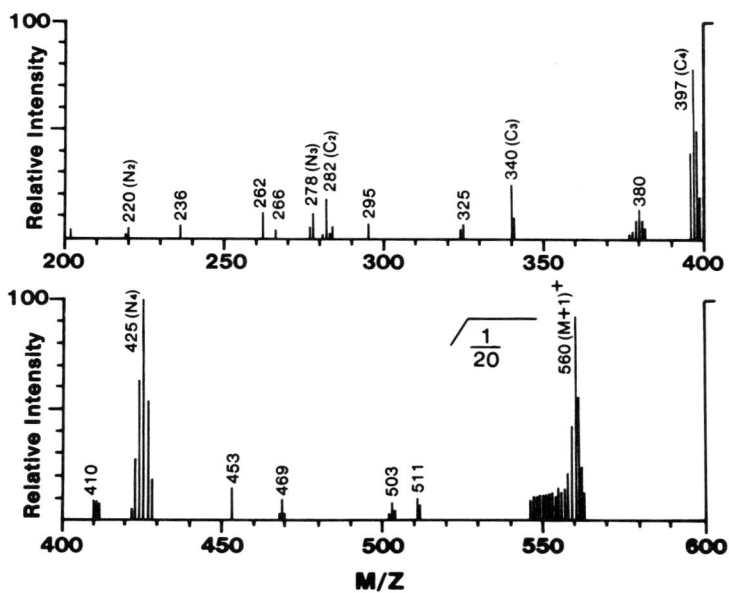

Fig.7.3. FAB-CAD-B/E mass spectrum of $^{18}O_2$-LE (11).

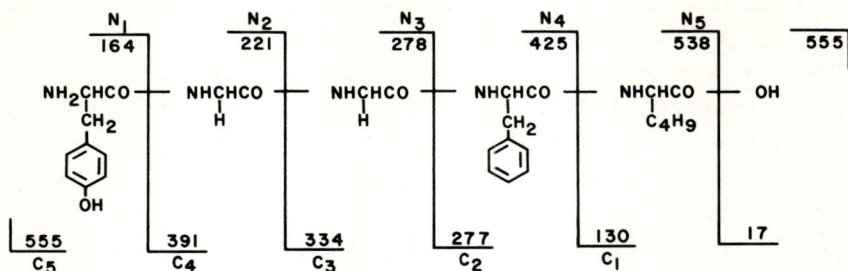

Fig.7.4. Fragmentation pattern of leucine enkephalin.

Based on a large body of MS experience for quantification of a variety of biologically important compounds (20,21), the most appropriate internal standard for measurement of an endogenous peptide is a stable isotope-incorporated peptide. This type of compound possesses the most closely-related physico-chemical characteristics in terms of extraction efficiency; RP-HPLC retention time; and MS ionization efficiency, fragmentation, and mass. Because of the presence of several naturally-occurring multi-isotopic elements (^{13}C, ^{2}H, ^{15}N, ^{17}O, ^{18}O, ^{33}S, ^{34}S) always present in any organic compound, the $(M+H)^+$ ion of a peptide internal standard must be increased at least four mass units relative to that $(M+H)^+$ of the peptide of interest to avoid any overlap of the isotope peaks from the two $(M+H)^+$ peaks.

Deuterium-containing amino acid residues can be readily incorporated into the peptide synthesis scheme to cause that needed increase in mass. Solid phase peptide synthetic methods, coupled with efficient HPLC purification methods, make this synthesis a possible option to produce peptide internal standards. However, MS measurement of endogenous peptides is then severely hampered by limited availability of appropriate synthetic capabilities. Furthermore, because of the rapidly increasing number of neuropeptides of biological interest which are being discovered, have biological activity, and need to be quantified, a laboratory would require a constantly increasing synthetic flexibility and speed for both production and purification of a large number of peptide internal standards.

Fast and facile incorporation of a stable isotope, for example ^{18}O, into the free carboxyl group of any internal amino acid residue provides the maximum possible flexibility for synthesis of stable isotope-incorporated peptide internal standards. Furthermore, the ^{18}O-incorporated peptide internal standards and endogenous ^{16}O-peptides possess virtually equivalent RP-HPLC and MS properties. While it is known that stable isotope-labeled compounds do have slightly different hydrophobicity values (22) [for example, in the present case, a peptide having either an ^{18}O or ^{16}O atom incorporated into the molecule], the larger oxygen atom (^{18}O) vs. deuterium atom (^{2}H) confers minimal differential hydrophobicity properties to the peptide.

It is a simple and straightforward task to synthesize this type of peptide standard by utilizing $H_2{}^{18}O$ as a convenient commercially available source for the stable isotope. Either acid-catalyzed or porcine esterase II-catalyzed (23) exchange reactions can be utilized (7). Furthermore, a combination of the two catalysts is possible. Figure 7.5 outlines the reactions used to incorporate ^{18}O atoms into LE using Fischer esterification, HPLC separation of the products, and porcine esterase recovery of the incorporated ^{18}O atoms. Data in Figure 7.5 show that, following Fisher esterification, three HPLC peaks are found due to;

(i) peptide bond hydrolysis;

(ii) direct incorporation of zero, one, or two oxygen atoms into the peptide carboxyl group;

(iii) methyl esterification, with incorporation of zero or one ^{18}O atoms. In the latter case, exchange is with either oxygen atom.

The rather expensive ^{18}O stable isotope is recovered conveniently by using porcine esterase II where again, zero, one, or two oxygen atoms are incorporated. For LE, the total yield of $^{18}O_2$ LE is 56%.

Figure 7.6 shows the reaction kinetics of ^{18}O-incorporation into the C-terminus of the opioid pentapeptide LE (upper) and the LE methyl ester (lower). In the upper panel, the reaction kinetics demonstrate that the $^{16}O_2$ species decreases rapidly, and is practically gone within approximately three days. The mixed isotope species $^{16}O{}^{18}O$ maximizes within one day, and then decreases smoothly. The $^{18}O_2$ species increases steadily and, after four days, levels out to approximately 90%. In the lower panel containing the methyl ester data, two complementary curves demonstrate incorporation of the ^{18}O species and the corresponding decrease of the ^{16}O species.

The two FAB-CAD-B/E scan mass spectra presented in Figures 7.2 and 7.3 confirm unambiguously that the two ^{18}O atoms are incorporated into only the carboxy terminal group of LE and not into any peptide amide bond or tyrosine hydroxy sidechain. Only the C_i-containing amino acid sequence-determining ions shift in the latter mass spectrum.

^{18}O-incorporated peptide internal standards were used first for measurement of endogenous ME and LE in brain and tooth pulp tissue and CSF extracts. Several tissues were studied in an effort to elucidate basic molecular mechanisms involved in noception (pain). Caudate nucleus,

LEUCINE ENKEPHALIN (500 μg)

MeOH · HCL, H$_2$ ^{18}O
HPLC

Peptide Bond
Hydrolysis
22%

^{18}O
LE—C—^{16}OH
^{18}O
LE—C—^{16}OH

(28 μg)

(200 μg)
40%

^{18}O
LE—C—^{18}OH (172 μg)

^{16}O
LE—C—^{16}OH

(190 μg)
38%

^{18}O
LE—C—^{16}OCH$_3$ (165 μg)

^{16}O
LE—C—^{18}OCH$_3$ (25 μg)

Porcine Esterase II

^{18}O
LE—C—^{16}OH

^{16}O
LE—C—^{18}OH

(20 μg)

Total Yield of ^{18}O$_2$—LE

$$\frac{172 + 108}{500} = 56\%$$

^{18}O
LE—C—^{18}OH   (108 μg)

^{16}O
LE—C—^{16}OH

Unhydrolyzed Esters (4 μg)

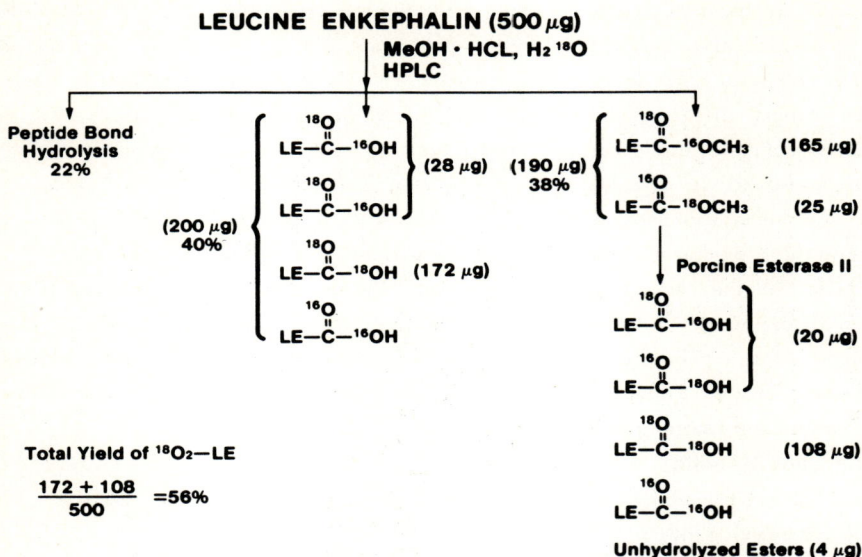

Fig. 7.5.   Chemical incorporation of ^{18}O into LE (7).

hypothalamus, pituitary, tooth pulp, spinal cord, thalamus, and CSF were
studied (24).

7.6 TYPES OF MASS SPECTROMETRY MEASUREMENT MODES

At first, FD MS was used to produce the (M+H)$^{+}$ ion of the peptide
being measured, and quantitative analytical data were obtained using the
SIM mode.   While this measurement procedure works well and provided the
first endogenous peptide measurements (5), FD MS has certain inherent
practical limitations.   For example, it takes a very high level of technical
expertise to fabricate appropriate FD emitters for measurement of peptides
derived from biologic extracts. Second, depending on the biologic tissue
studied and the purification processes used prior to MS, a certain percentage
of the fragile FD emitters will unavoidably break. Even with excellent
technical expertise and years of experience, this breakage problem can be
minimized, but probably never completely eliminated. Third, the (M+H)$^{+}$ ion
current produced with FD is weak and highly fluctuating. While a microcomputer
(Apple II) interfaced to both the emitter heating control and multiplier
output amplifier circuitry of the MS instrument partially overcomes these
technical problems (14), the problems are not completely eliminated.

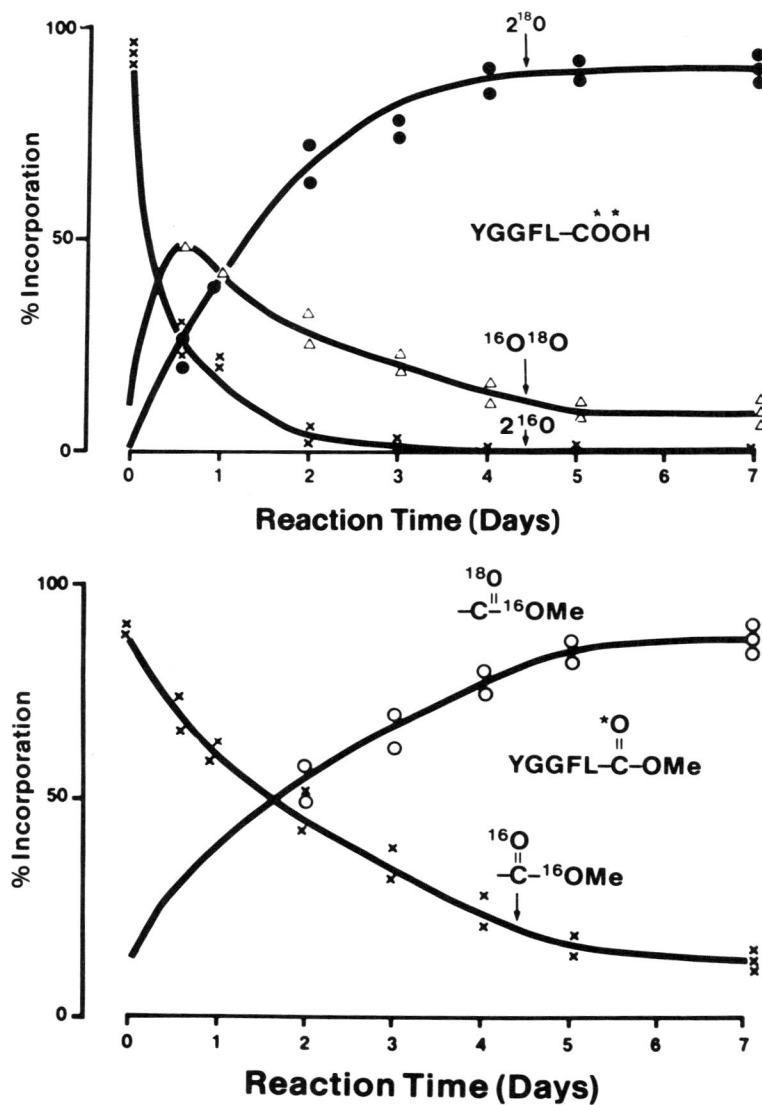

Fig. 7.6. Kinetics of the incorporation of ^{18}O into LE (7). Upper panel-incorporation of two ^{18}O atoms into LE. Lower panel-incorporation of one ^{18}O atom into LE methyl ester.

FAB MS production of the $(M+H)^+$ ion of a peptide overcomes most of the technical problems which have been experienced for FD processes. However, because of the nonvolatile matrix (generally, glycerol) needed to constantly present a fresh molecular surface of peptides to the fast atom beam, "effective resolution" of the MS analysis procedure must be increased in either one of two ways. In the future, perhaps both methods can be utilized or combined. Either the mass resolution must be increased to one part in 20,000- 30,000 to be able to select only the accurate mass of the specific $(M+H)^+$ ion of the peptide of interest [for example, $(M+H)^+$ of leucine enkephalin, $C_{28} H_{38} N_5 O_7 = 556.2771$] or else a linked-field scan, with or without CAD, must be used to increase the "structural resolution" of the procedure and to produce a unique amino acid sequence-determining ion from the selected $[M+H]^+$ (15). The latter methodology is the preferred option for peptide measurement because increasing the instrumental mass resolution to monitor the accurate mass of an endogenous peptide still does not confer sufficient molecular specificity (structure) to the analytical measurement (see above). On the other hand, use of the combination CAD-B/E scan or SIM processes extends the effective structural resolution of both the MS and the HPLC systems, and also increases the analyst's confidence of the measurement process, because now an amino acid sequence-determining ion is selected to form the basis of measurement. This increased specificity derives from the fact that a mass spectrum is obtained from one selected ion in a FAB mass spectrum, and an amino acid sequence-determining product ion of this selected precursor ion is chosen for quantitation.

Optimal molecular specificity for endogenous peptide measurement and ease of production of $(M+H)^+$ are both achieved by using the novel mass spectrometric combination of FAB-CAD-linked-field (B/E and B'/E')-SIM coupled with a microcomputer (14).

Modifications to both the MS instrument and microcomputer permit alternation (sec $^{-1}$) between two selected linked-field (B/E) values - namely, B/E for the $(M+H)^+$ of the endogenous peptide of interest and B'/E' for the $(M+H)^+$ of the stable-isotope incorporated peptide internal standard which is located generally four mass units higher.

7.7 EXAMPLES OF ANALYTICAL MEASUREMENTS OF ENDOGENOUS ENKEPHALIN PEPTIDES

This section discusses two experimental topics. On one hand, techniques involved in the construction of appropriate calibration curves are discussed. On the other hand, a variety of selected HPLC chromatograms is presented to visually demonstrate the range of metabolic profiles of the peptide-rich fractions which are extracted from several biologic tissues and CSF fluid.

7.7.1 Construction of Calibration Curve

The method of analysis employing the FAB-CAD-B/E-B'/E'-SIM-microcomputer measurement mode must demonstrate a linear response over the range of concentrations corresponding to endogenous peptides. The use of the stable isotope-incorporated peptide internal standard with this novel analytical measurement mode is important to demonstrate linearity and to overcome biologic matrix effects. Primary data from one of the accelerating voltage alternation experiments are shown (Figure 7.7) for LE (two micrograms) versus O^{18}-LE (two micrograms). The monitored peak is mass 336 in the former case and 340 in the latter case. This selected ion corresponds to the tripeptide sequence -GFL which derives from the CAD-B/E scan mode. This amino acid sequence-determining ion is a product ion arising only from the $(M+H)^+$ ion (556) of LE - a most significant experimental fact that rigidly maintains the molecular specificity of the analytical measurement. In the oscillographic trace (Figure 7.7), the ^{18}O-species is clearly indicated when the higher mass (M_S) is being monitored because a 60 Hz signal is superimposed on a lower trace. The lighter mass, 336, due to endogenous LE, is denoted by M_L. The microcomputer accepts a large number of signals to optimize ion statistics. Integrated areas are plotted and the ratio of the known amounts of the ^{16}O and the ^{18}O species are plotted. Figure 7.8 contains the calibration lines for both the ME (upper) and LE (lower) experiments. Both lines intercept at, or very near, the origin and have correlation coefficients near unity. The statistical parameters for the two best-fit straight-lines are, for ME: $y = 1.09x + 0.06$, $r^2 = 0.999$ and for LE: $y = 0.69x$, $r^2 = 0.995$.

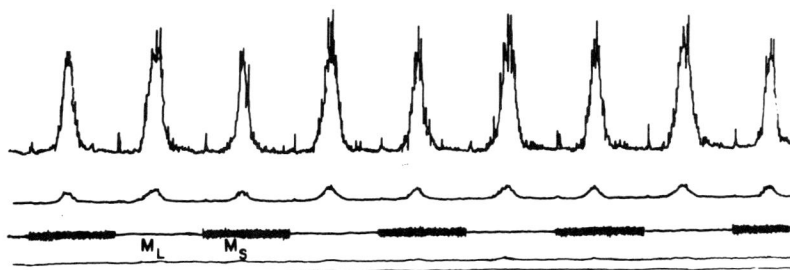

Fig. 7.7. Trace of analog data which is obtained during MS measurement of LE (M_L, 336) and $^{18}O_2$-LE (M_S, 340); (24).

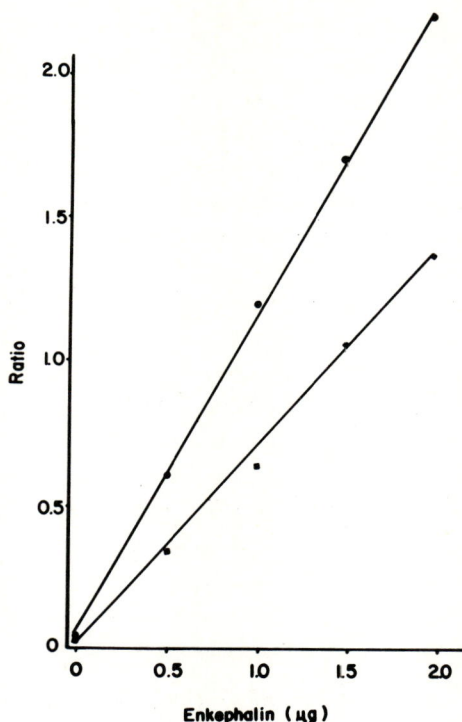

Fig. 7.8. Integrated MS peak areas vs. amount of ME (upper line) and LE (lower line); (24).

7.7.2 Hypothalamus

The RP-HPLC for canine hypothalamus tissue (two grams) is shown in Figure 4.5 and the chromatogram of sample spiked with the two enkephalin peptide internal standards $^{18}O_2$-ME and $^{18}O_2$-LE is shown in Figure 4.6. The two arrows indicate the known retention times of synthetic ME and LE, respectively. It should be noted that, at the UV wavelength being monitored (200 nm), the RP-HPLC chromatogram of a hypothalamic peptide-rich fraction is relatively clean of UV-absorbing material at that noted sensitivity level. It is quite important to remember that of course, a wealth of other biologically active RRA, BA, or RIA responding material may, and generally does, coelute on this chromatogram at the indicated at other retention times (25,26). It is experimentally observed that, at a UV detection sensitivity level of 0.1 AUFS, microgram amounts of peptides are detected, whereas RRA, RIA, and BA are capable of detecting ng and pg amounts.

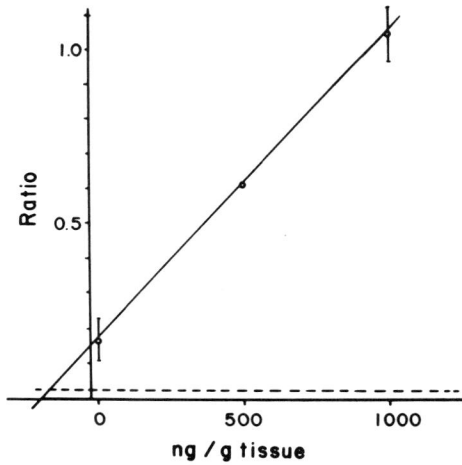

Fig.7.9. Plot of FAB-CAD-B/E-B'/E'-SIM data for measurement of LE in a hypothalamus extract data (24).

TABLE 7.1. Mass spectrometric measurement of enkephalins in canine tissue extracts (ng^{-1}) and CSF $(ng\ ml^{-1})$ (24).

SOURCE	ENKEPHALIN	AMOUNT	METHOD*
Hypothalamus	LE	170	1
Cerebrospinal Fluid	LE	44	1
Pituitary			
● Anterior	LE	70	2
	ME	2,950	2
● Posterior	LE	2	2
	ME	3,760	2
Caudate Nucleus	LE	1,500	2
Tooth Pulp**	LE	30	2
	ME	179	2
Tooth Pulp			
● Control	LE	20	2
	ME	487	2
● Electrostimulated	LE	45	2
	ME	390	2

*Method 1: FDMS-SIM-Microcomputer;
      2: FAB-CAD-B/E-B'/E'-SIM-Microcomputer
**Unstimulated pooled tissue from five dogs

While experimental experience indicates that the RP-HPLC resolution of the two enkephalin peaks may be increased by alterations of several experimental parameters including recycling and/or change of the buffer and/or organic modifier, flow rate, temperature, etc., it must be remembered that in the novel mode of MS analysis, the detector is not limited to only UV absorption, but rather, utilizes a unique amino acid sequence-determining ion arising from a peptide eluting at one selected retention time. In this type of measurement mode, virtually all chromatographic and chemical background noise disappears. The plot of LE and the ^{18}O-LE internal standard (M+H)$^+$ ratios for this hypothalamus data is shown in Figure 7.9, where intersection of the signal and the noise level (dotted line) corresponds to 170 ng leucine enkephalin g^{-1} hypothalamus tissue. Table 7.1 lists the amounts of enkephalin determined in a variety of canine tissues.

7.7.3 Cerebrospinal Fluid

Figure 4.7 contains the RP-HPLC of canine CSF alone (2 ml) and Figure 4.8 the corresponding chromatogram of CSF which is spiked with the two peptide internal standards. Both LE (44 ng ml^{-1}) and ME (84 ng ml^{-1}) are measured in this extract.

7.7.4 Pituitary

A number of canine pituitaries [10] was accumulated and neuroanatomically separated into the anterior (1.9 g. total wet weight) and posterior (0.44 g total wet weight) portions. The tissue is homogenized in acetic acid (1 M) and divided into three equal samples. Figure 7.10 contains the RP-HPLC of the anterior pituitary tissue extract alone, and Figure 7.11 the anterior pituitary extract plus two peptide internal standards. Figures 7.12 and 7.13 contain the two corresponding RP-HPLC chromatograms for the posterior pituitary extraction. Table 7.1 contains the amounts of the enkephalins found in both the anterior (LE = 70; ME = 2,950 ng g^{-1}) and posterior (LE = 2; ME = 3,760 ng g^{-1}) pituitary extracts.

7.7.5 Caudate Nucleus

FD-MS methods have been used to measure the amount of endogenous enkephalin in canine nucleus tissue extracts (17, 18). The amount of endogenous enkephalin in the canine caudate nucleus tissue extract is 190 ng of ME and 1,500 ng LE g^{-1} tissue. This measurement represents one of the few times where the concentration of LE exceeds that of ME.

7.7.6 Tooth Pulp

Tooth pulp tissue is collected from four animals and pooled (four teeth from each animal; total = 16). Canine tooth pulp RP-HPLC chromatograms have been published (16). The endogenous amount of enkephalin for pooled tooth pulp tissue is 179 ng ME and 30 ng LE g^{-1} tooth pulp tissue, respectively.

Fig. 7.10. RP-HPLC chromatogram of the peptide-rich fraction extracted from anterior pituitary tissue. Arrows denote retention time of ME and LE, respectively (24).

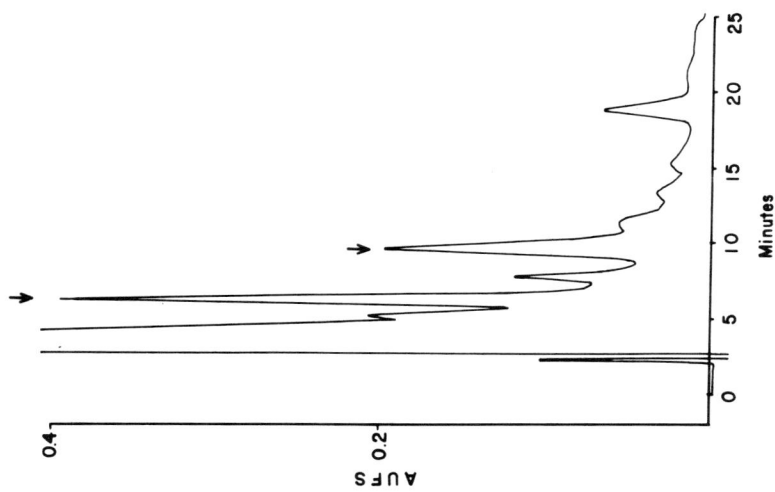

Fig. 7.11. RP-HPLC chromatogram of the peptide-rich fraction extracted from anterior pituitary tissue which has been spiked with ^{18}O-ME and ^{18}O-LE peptide internal standards (arrows, respectively); (29).

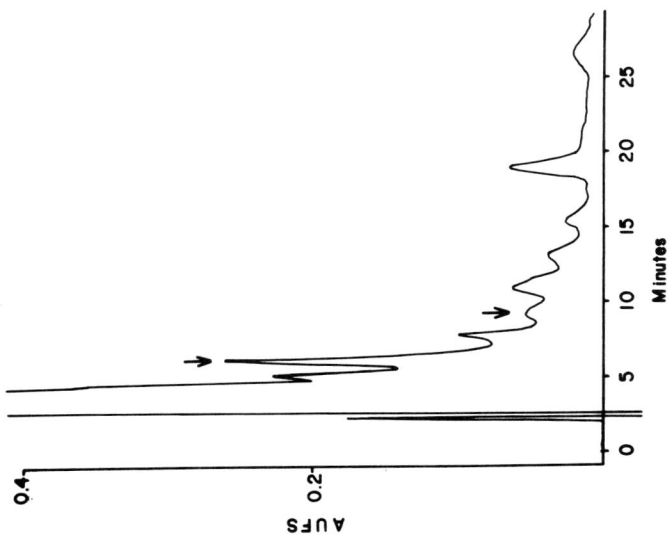

196

Fig. 7.12 RP-HPLC chromatogram of the peptide-rich fraction extracted from posterior pituitary tissue (24).

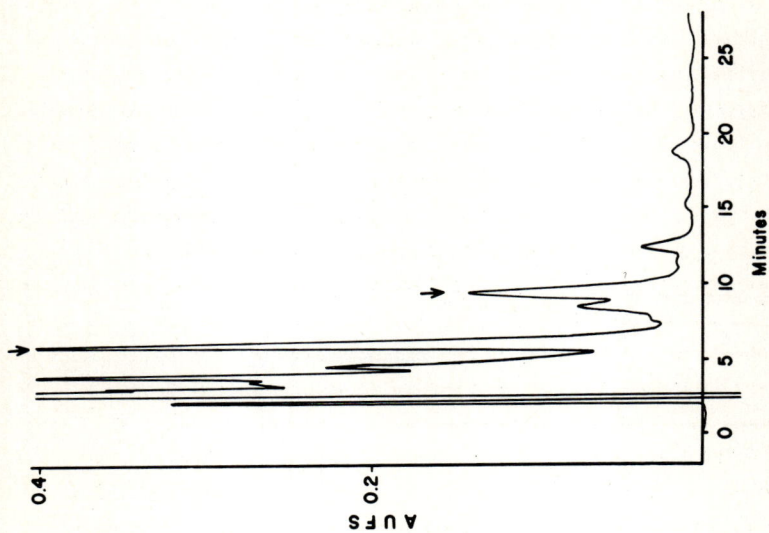

Fig. 7.13. RP-HPLC chromatogram of the peptide-rich fraction extracted from posterior pituitary tissue which has been spiked with ^{18}O-ME and ^{18}O-LE peptide internal standards (arrows, respectively); (29).

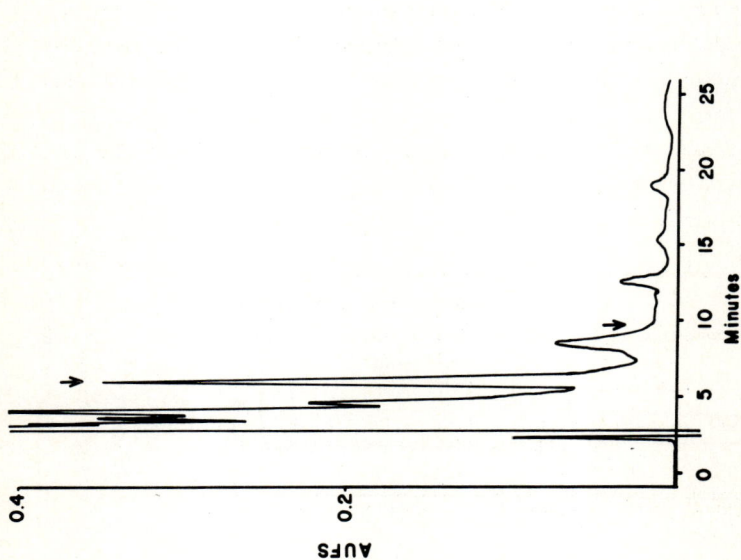

7.7.7 Electrostimulated Tooth Pulp

In a study to determine whether electrical stimulation affects endogenous enkephalin concentrations, three animals served as control and three for electrostimulation experiments. The current working hypothesis underlying this type of physiological study is that three peptidergic pathways (endorphinergic, dynorphinergic, and enkephalinergic) are available to a cell to maintain a dynamic homeostatic relationship and to deal with noxious stimuli. The three peptidergic pathways are composed of large and intermediate precursor peptides, the opioid oligopeptide, and metabolites. Noxious stimuli are hypothesized to activate the peptidergic pathways and individual opioid peptides may have decreased concentrations following stimulation.

Table 7.2 collects the individual analytical measurements for this type of study. It is noted that animal 110 is exsanguinated before stimulation and relatively less tissue (0.528g) was obtained, but the values for ME and LE are considered to be valid. Also, ME data for animals 112 and 110 seem to be proportionately higher, due to perhaps a lower level of internal standard in the MS measurement, even though the corresponding RP-HPLC data appear acceptable. These two questions notwithstanding, it is noted (Table 7.1) that electrostimulation significantly decreased by 20% the amount of endogenous ME while doubling the amount of LE.

A general overall trend noted, is that both the opioid pentapeptides ME and LE are altered upon electrostimulation (26-28). Electrostimulation is performed to elucidate those molecular mechanisms which occur during a physiologically stressful situation. These preliminary data indicate that the three peptidergic pathways (dynorphinergic, endorphinergic, enkephalingeric)

TABLE 7.2. Individual mass spectrometric measurements of enkephalins in canine tooth pulp extracts (24).

(ng enkephalin g^{-1} tissue)

	Dog	ME	LE	Wet Weight (g)
Control, Unstimulated	112	750	30	1.714
	114	90	9	2.090
	127	620	20	1.071
Stimulated	110*	690	120	0.528
	115	390	10	1.637
	134	90	6	1.711

*Exsanguination begun before electrostimulation.

198

may be mobilized in the following sequence: Large precursor ────▶intermediate
precursor(s) ──────▶ pentapeptide(s) ──────▶ inactive metabolites (see
Chapter 2). On one hand, there may be a naturally-occurring pool of
pentapeptides which is electrostimulated towards metabolism or, on the other
hand, the entire metabolic scheme noted above may be stimulated to produce a
lowered endogenous amount of each constituent peptide. Other human and in
vivo dynamic studies are needed to resolve that question.

7.7.8 Conclusions

Several significant conclusions are derived from the experiments
reported in this section. A fast and facile method of tissue sample
acquisition and procurement from the canine animal model is described. This
study demonstrates that rapid freezing of tissue is needed in a fashion which
is similar to that described in the discovery of 19OH-PGE$_2$ compounds in human
seminal fluid (29-31). Rapid freezing avoids, or at least minimizes, metabolic
and chemical interconversions and also enhances the possibility of measuring
only those endogenous target compounds and not artifacts or chemical/enzymic
products. The need for an internal standard for an MS analytical measurement
is demonstrated by other workers (20, 21) to overcome the experimentally
rather ill-defined, yet very real, biological matrix effects. Stable
isotope-incorporated peptides are the most appropriate internal standards for
measurement of endogenous peptides. An internal standard is added as soon as
possible after tissue acquisition and before homogenization in the separation
scheme as a means to accurately represent the endogenous amount of peptide
and to also provide sufficient time for equilibration (20) of the exogenous
and endogenous peptides.

The most significant experimental parameter of any analytical
measurement of a biological compound is the molecular specificity of that
measurement; namely, is the compound one thinks is being measured the
compound that is actually being measured? This concept of specificity is easy
to state, but experimentally rather difficult to unambiguously prove. One
author calls this experimental phenomenon the "chromatographic uncertainty
principle" (32). Many other assay methods are generally utilized because of
their relative ease, low cost, high speed, high sensitivity, and putatively
high molecular specificity. For example, chromatography, color reactions,
enzymatic reactions, HPLC, BA, RRA, and RIA are the detection methods which
are most-used in laboratories around the world. However, this author states
unequivocally that the molecular specificity of all of the above measurement
procedures is insufficient for unambiguous structural proof during an
analytical measurement. Of course, the non-MS assay methods listed above will

always be used, but investigators must at least be aware of and state the limitations of any statements which they may make regarding structure. Only one measurement process, namely MS, offers unambiguous molecular specificity. On one hand, the ability to produce the $(M+H)^+$ ion of a biologically important peptide is a signal advancement in the measurement of endogenous peptides. But even the use of this parameter, $(M+H)^+$, while significantly increasing specificity, does not confer unambiguous molecular specificity to that measurement. The only analytical method which is currently available and which uses MS to provide maximum molecular specificity is to select an amino acid sequence-determining ion from the FAB-produced $(M+H)^+$ ion, either by unimolecular or CAD processes, and then to collect by a linked-field scan only one unique amino acid sequence-determining ion. Furthermore, use of a stable isotope-incorporated internal standard which corresponds to the same peptide as that which is being measured, additionally substantiates the molecular specificity of the analysis. In the study discussed above, the C-terminal tripeptide sequence ions -GFL from LE and -GFM from ME are selected for monitoring, and the analytic measurement of endogenous peptides is based upon those two ions and their corresponding ^{18}O-internal standards.

The detection sensitivity of the novel MS process described in this section is quite encouraging for analytical measurement of endogenous peptides in most biological tissues and fluids. For example, enkephalin peptides in canine caudate nucleus tissue extracts are measured at the 200-400 ppb level. The current instrumental limitation corresponds to the 30 ppb level. It is encouraging to realize that several significant instrumental increases (10-100x) in the detection limits are forthcoming.

The peptide measurements listed in Table 7.1 include endogenous amounts of ME and LE pentapeptides extracted from a variety of biological sources including hypothalamus, CSF, anterior and posterior pituitary, caudate nucleus, and tooth pulp (pooled and electrostimulated). It is important to undertake this type of analytical/physiological study within one laboratory to ensure quality control over all experimental manipulations which range from the live animal model through exsanguination, tissue procurement, homogenization, chromatography, MS, and data analysis. Inter-animal biologic variations are observed, and it may be possible to have one animal serve as its own control.

7.8 COMPARISON OF RIA AND MS VALUES

It is important to compare literature values for RIA measurements of enkephalin with the newer MS procedure. For example, in two separate canine studies using MS techniques, measurement values for LE were 423 and 351 ng, while for the hypothalamus, 268 and 240 ng g^{-1} were found. These MS data are generally compatible with RIA data (33-37). The wide range (37-fold) of RIA values can be rationalized by several technical considerations such as: differences in antibodies used for RIA; handling of the tissue (time to acquire); temperature; peptidase, synthetase, and enkephalinase activity; species differences; whether or not chromatography was used before RIA; age of animal; and possibly the "emotional state" of the animal at the time of sacrifice (38).

It is realized that MS instrumentation is sophisticated to operate, costly to purchase, and is not readily available to the majority of the laboratories which are performing neuropeptide research. It is interesting to note however, that whenever factors such as the amortization of the initial capital used to purchase the MS instrument, time required per analysis, cost per analysis, number of personnel shifts used per day, etc. are included in a cost analysis, a FAB-CAD-B/E-B'/E'-SIM-microcomputer analysis of endogenous peptides is directly competitive with RIA analysis (ca. $40-50/sample).

In any event, wherever possible, other assay methods used for peptide quantification should be calibrated with the structural certainty that only MS can offer. It is clear that the combination of HPLC and MS will not replace RIA due to the advantages that the latter method provides in terms of competitive cost, high speed, and high sensitivity, but at least an independent and primary assay method is now available to the neurochemist.

7.9   SCHEME FOR ENDOGENOUS PEPTIDE AMOUNTS

Figure 7.14 contains the scheme used to calculate the amount of endogenous peptide in a biologic extract. A known amount of a stable isotope-incorporated peptide internal standard, generally at the ppm level, is added to the extract, and the sample is divided into two parts. Both parts are subjected to Sep-Pak, analytical HPLC, and MS measurement, but one part has an additional amount added of the target peptide, generally 200 ppb. Solution of the two equations for the two unknowns yields x, the amount of endogenous peptide.

Two other methods discussed below are used to answer the two questions: a) are only two measurements (spiked, unspiked) sufficient for analysis of endogenous peptides? and b) what is the limit of the detection sensitivity?

During measurement of pmol amounts of LE in canine spinal cord tissue, a scheme was devised to construct the calibration curve for measurement of that peptide in a series of biologic extract samples by the standard addition

TISSUE

I.S. $(1\,\mu g/g)$

| 50% | 50% |

Sep-Pak (RP)
RP-HPLC
MS

LE Spike $(200\,ng/g)$
Sep-Pak (RP)
RP HPLC
MS

$ax = R_1$

$a\,(x+200) = R_2$

$$a = \frac{R_2 - R_1}{200}$$

$$x \cdot 200 \frac{R_1}{R_2 - R_1}$$

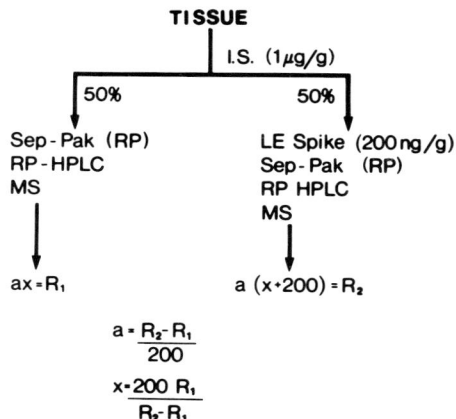

Fig.7.14. Method to calculate endogenous LE in spiked and unspiked thalamus tissue extracts. Slope = a, and R is ratio of integrated areas for spiked (R_2) and unspiked (R_1) samples.

method. Spinal cord tissue (ten g) is divided into five equal portions and increasing amounts of synthetic LE are added to four portions of the tissue extracts (100, 200, 400, 1000 ng g^{-1} tissue). The straight line obtained by this experimental procedure has a best-fit equation of $y = 0.00237\ x + 0.35$ and a correlation coefficient (r^2) of 0.99. This staight line intercepts the abscissa at a value corresponding to 70 ng (127 pmol) of endogenous LE in spinal cord tissue.

A second experiment was designed to test the question of whether this analytical method is the most statistically significant approach to this type of unique peptide measurement, as compared to an MS experiment comparing only two measurements - endogenous peptide alone and endogenous peptide plus a single addition of a known amount of synthetic peptide. Towards that end, incremental amounts of synthetic LE are added to the sample extract and the data are projected towards zero addition. Data from this experiment confirm that the previous method of using only two measurements is sufficient for determination of neuropeptides in brain extracts at the part per billion (pmol) level (10).

A calibration curve was determined to calculate the level of sensitivity achievable with the combination of HPLC separation and FD MS measurement of synthetic solutions of opioid peptides. Four samples were prepared, where each sample contained 300 ng of the internal standard, [2]ala-leucine enkephalin.

To each one of these four samples, a successively lower amount (300, 100, 30, and 10 ng) of LE was added to produce solutions with ratios of 1:1, 3:1, 10:1, and 30:1 for LE to internal standard, respectively. Down to 30 ng of LE is readily measured by this method, and this amount corresponds to 60 pmol. The straight line relating the ratio of LE to internal standard versus the weight of ^{2}ala-leu (in ng) has a correlation coefficient (r^2) of 0.999, and the best-fit equation is y = 0.00284x + 0.005.

7.10 METABOLIC PROFILES OF PEPTIDES IN TOOTH PULP

It is important to be able to discern what effect, if any, a physiological experiment will have on the three peptidergic pathways discussed above. Towards that end, metabolic profiles were obtained for peptides contained in canine tooth pulp extracts (39). Several immobilized proteolytic enzymes (chymotrypsin, trypsin, and carboxypeptidase A) were utilized to determine which HPLC peaks derive specifically from peptides. Results from treatment of these tooth pulp extracts with immobilized enzymes demonstrate clearly that virtually all peaks in this HPLC system are due to peptides, because the larger peptides (higher hydrophobicity, later-eluting peaks) are destroyed by enzymes to produce smaller peptides (lower hydrophobicity, early-eluting peaks).

To demonstrate this procedure, Figure 7.15 contains the RP-HPLC phase gradient chromatogram of a control tooth pulp tissue. This control sample is subjected to trypsin enzyme treatment. The experimental conditions for this chromatography are: one microbondapak C_{18} column; 200 nm; 1.5 ml min^{-1} flow rate; 0.1 AUFS. The gradient profile is noted by the line on the chromatogram, where the shaded peaks indicate those peaks which disappear following trypsin treatment. Trypsin (TPCK-treated) on 4% beaded agarose is used. Lyophilized tooth pulp peptide extract (100 ml sample, 250 dry solid equivalant to 70 ml sample tissue) in TEAP buffer (0.06 <u>M</u>, pH 2.12) is adjusted to pH 8.5 with one-two microliter of concentrated ammonium hydroxide. Sample is stirred one hour with the appropriate peptidase (1.5 units trypsin). One-half of the sample is analyzed by RP-HPLC. Figure 7.15 contains that RP-HPLC chromatogram of an unhydrolyzed tooth pulp extract, and demonstrates that most peptides elute at a higher hydrophobicity (percentage acetonitrile organic modifier). This RP-HPLC represents for the first time a metabolic profile of peptides found in a tooth pulp extract. Shaded areas in Figure 7.15 indicate those peaks which will disappear after treatment with trypsin. Figure 7.16 contains the RP-HPLC gradient chromatogram of a corresponding tooth pulp extract in Figure 7.15, but now following

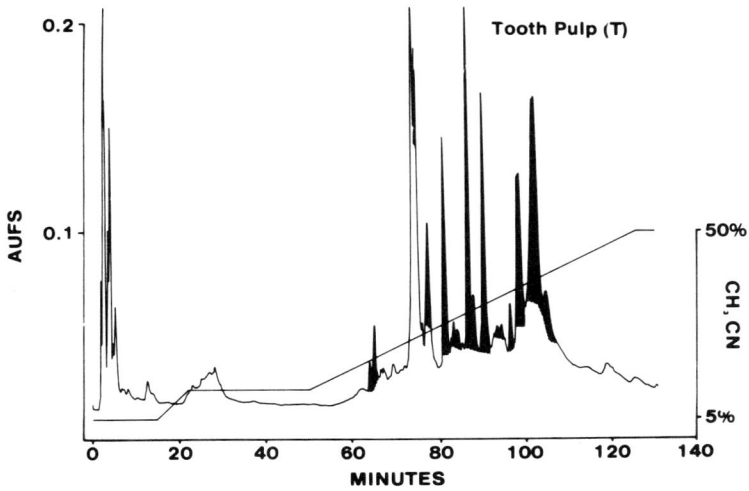

Fig.7.15. Gradient RP-HPLC chromatogram of the peptide-rich fraction from control canine tooth pulp. Shaded peaks indicate trypsin-sensitive fractions (39).

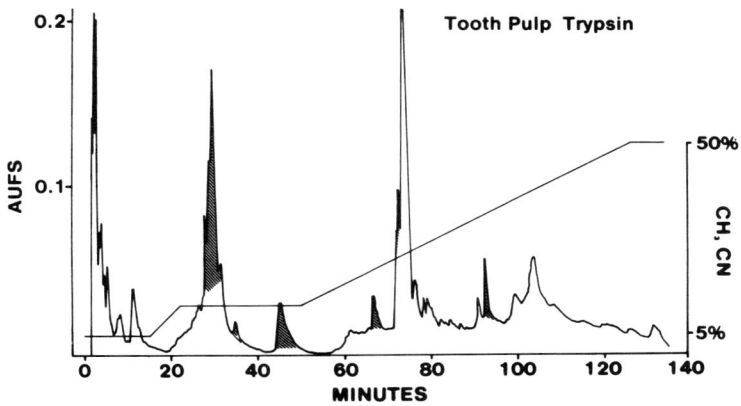

Fig.7.16. Gradient RP-HPLC chromatogram of trypsin-treated peptide-rich fraction from canine tooth pulp extracts. Shaded peaks indicate new peaks which are formed after trypsinolysis (39).

treatment with trypsin. Shaded areas of Fig. 7.16 indicate those peaks which appear after trypsin treatment, and derive from larger peptides which contain a trypsin-sensitive basic amino acid residue (Arg or Lys).

This type of study is a component in a larger research program focusing on monitoring and quantifying the enkephalin-, dynorphin-, and endorphin-related peptides in biologic extracts including brain, tooth pulp tissue, and CSF. Alpha-chymotrypsin on cellulose, carboxypeptidase A on agarose, and trypsin on 4% agarose are commercially available immobilized enzymes. In each enzymolysis study, an HPLC gradient (5-50%) of organic modifier (acetonitrile) is utilized to elute larger proteins and peptides. Following enzymolysis, nearly all of the higher molecular weight species shift to a lower retention time, reflecting peptide bond cleavage and production of smaller peptides having a decreased hydrophobicity. For extracted peptides which may have too large of a molecular weight (for example, beta-endorphin has a MW=3,624) for the MS analysis of the intact peptide using the older mass spectrometers of more limited mass range capability, immobilized enzymolysis treatment of the peptide, followed by gradient HPLC elution and perhaps isocratic purification and salt removal of individual fractions by RP-HPLC will provide enzymic peptide fractions which have molecular weights that are amenable to MS quantification methods. Recently, however, FAB MS was used to study intact proinsulin (MW=9,000) (40), and this type of newer high mass MS instrumentation may decrease the need for these extensive enzymolysis studies.

Another use for RP-HPLC/enzymic characterization of peptides extracted from tooth pulp tissue involves total molecular characterization of the RP-HPLC peaks (see Chapter 5). Towards that end, peaks to be characterized first are those HPLC fractions which display biologic, immunoreactive, or opiate receptor activity, or those HPLC peaks which change concentrations following physiological alterations. A "screen" utilizing either RRA or RIA antibodies (raised against ME, LE, beta-endorphin, and other neuropeptides) may also be utilized, where the former is more efficient for this purpose due to its wider range of "screening" reactivity. Current experiments are being done using a gradient RP-HPLC chromatographic separation using a canine limbic system synaptosome fraction (25) as a RRA detector for the HPLC (see Section 5.4.8).

7.11 PEPTIDE EXTRACTION AT NEUTRAL pH

Previous extraction schemes with [2]ala-leucine enkephalin as internal standard utilized homogenization in acetic acid (1 M), followed by protein precipitation with acidified acetone. However, with the recently-developed stable isotope-incorporated peptide internal standard, back-exchange of ^{18}O for ^{16}O atoms occurs at acidic pH values. Therefore, another extraction scheme at neutral pH was devised for efficient extraction of the endogenous peptides from a biologic tissue extract. Homogenization is done in ethanol, and the C-18 Sep-Pak effluent of the peptide-rich fraction is dissolved in Tris buffer (50 mM, pH 7.5). Four fractions are eluted: fraction 1 (4 ml sample solvent); fraction 2 (4 ml Tris buffer); fraction 3 (4 ml of methanol: Tris buffer = 10:90); and fraction 4 (2 ml of methanol:Tris buffer = 80:20). Studies with radiolabeled peptides demonstrate that fraction 4 contains the majority of the peptides, and therefore that fraction is collected, lyophilized, and subjected to analytical RP-HPLC.

For the first time using this procedure, canine thalamus tissue was extracted and two opioid peptides measured in one experiment (7). Thalamus tissue (2g) was spiked with two micrograms each of ^{18}O-LE and ^{18}O-ME. Endogenous amounts of LE (62 ng) and ME (125 ng) were measured. This type of methodology is readily extended to other peptides.

7.12 COMPARISON OF FIELD DESORPTION AND FAST ATOM BOMBARDMENT MASS SPECTROMETRIC MEASUREMENT METHODS

This section describes the first, one-on-one comparison of FD versus FAB mass spectrometric measurements (12). Until recently, FD MS was the only available ionization method for studies of peptides. In general, FD methodology works well, and many analytical measurements of endogenous peptides were made utilizing FD MS of the $(M+H)^+$ ion in the SIM mode with a higher homolog of the peptide serving as internal standard.

A certain amount of unavoidable breakage of the delicate FD emitters was noted in the past several years of experience utilizing FD MS. While FD emitter breakage is not excessive when studying peptide standard solutions, unavoidable and real "biologic matrix effects" increase the amount of emitter breakage. Two further instrumental developments are available to both increase the reliability and convenience of the measurement process and to increase the molecular specificity of the analytical measurement. The development of FAB MS (See Chapter 6) provides an optional ionization mode which bypasses almost all of the technical problems associated with FD MS. Furthermore, the FAB measurement of the $(M+H)^+$ ion of a peptide in the SIM mode is performed more conveniently compared to FD MS. However, a higher order of molecular specificity is required which can be readily obtained by combining the FAB ionization mode with either unimolecular or CAD production of amino acid sequence-determining ions, followed by monitoring of selected product ions in the linked-field (B/E) scan mode. For example, amino acid sequence-determining information is obtained from nmol amounts of the chemically underivatized, biologically important undecapeptide substance P by combining FAB, CAD, and linked-field (B/E) scanning mass spectrometry (41). $(M+H)^+$ ions of the undecapeptide substance P are produced by FAB MS, accelerated to high translational energy (8 kV), and transit a collision chamber. CAD occurs in the first FFR of the instrument and amino acid sequence-determining ions are collected by scanning the electric and magnetic fields, keeping the ratio B/E constant. This scan mode collects all of the product ions produced from a selected precursor ion (generally $[M+H]^+$). In the B/E mode, the structurally informative precursor/product relationship between those ions produced during fragmentation of the $(M+H)^+$ ion is both firmly established and rigidly maintained. The $(M+H)^+$ ion produced by FAB MS is quite intense, while amino acid sequence-determining ions produced by CAD are generally only a few percent, relative to the abundance of the $(M+H)^+$. While at first this decreased intensity may seem to yield a drastic and unacceptable decrease in detection sensitivity, it must be realized that the linked-field scan mode greatly reduces the background noise level which is mainly due to chemical or biological noise. The remaining inherent instrumental electronic noise must still be considered. The decrease in the signal and noise levels in a linked-field scan fortunately combine to provide a signal-to-noise ratio which is comparable to that found for FD MS measurement of $(M+H)^+$ ion. This favorable signal-to-noise ratio enables measurement of endogenous peptides in biologic tissue extracts by MS methods by FAB-CAD-B/E.

A great increase in the molecular specificity of this novel analytical measurement process is achieved by selecting one unique amino acid sequence-determining ion to measure endogenous LE. An endogenous amount (451 pmol g^{-1} tissue) of LE in an extract of canine caudate nucleus is measured (12). For LE, the C-terminal tripeptide (-GFL) ion at mass 336 is selected as a unique, structurally unambiguous ion. The $^{18}O_2$-internal standard for LE produces a corresponding ion at mass 340 in the FAB-CAD-B/E-B'/E'-SIM mode. A plot of this type of data from a synthetic mixture of $^{18}O_2$- and $^{16}O_2$-LE [fixed amount of ^{16}O is measured, versus sequentially smaller amounts (500, 300, 100, 5, and 0 ng) of ^{18}O-LE] yields a straight line which, when corrected for background noise, passes through the origin. The best-fit equation of that line is $y = 0.00294x + 0.017$, with a correlation coefficient (r^2) equal to 0.99. FAB-CAD-B/E-B'/E'-SIM measurement of tissue extracts yielded a value for the endogenous amount of LE in canine caudate nucleus of 250 ng LE g^{-1} wet weight tissue corresponding to 451 pmol.

In a parallel experiment using the other half of the same tissue extract, FD MS in the SIM mode of the (M+H)$^{+}$ ion was used to quantify 160 ng g^{-1} of LE. The difference between the two analytical measurements reflects in part the difference in signal-to-noise ratios of the two ionization and instrumental analysis modes, where FD is found to have a higher signal-to-noise ratio. The amount of LE determined in this work compares to previously reported MS (423 and 351 ng g^{-1} of tissue) (17, 18) and RIA (210 ng g^{-1}) data (34).

FAB MS produces a steady and abundant (M+H)$^{+}$ ion current which lasts for minutes for endogenous amounts of peptides. FD, on the other hand, produces a spurious, rapidly fluctuating, and smaller ion current of peptides. While FD processes work well for peptide quantification, the more dependable technique of FAB [in the CAD-linked field (B/E) scanning mode] for peptide analysis is preferred. However, due to the high glycerol matrix background generally found with FAB, it is necessary to use CAD-linked field (B/E) scanning MS techniques to effectively increase the molecular specificity of the measurement. While this combination process takes advantage of the inherent signal-to-noise ratio, the alternate approach (increased mass resolution) would inordinately decrease sensitivity, and still not provide unambiguous specificity.

A chapter describing all of the recent developments discussed above has been recently published (15).

7.13 SUMMARY

For the first time, endogenous peptides are measured by MS methods in CSF, brain tissue, and tooth pulp tissue extracts. FAB produces an abundant (M+H)$^{+}$ which may or may not be subjected to CAD to either enhance or produce a unique amino acid sequence-determining ion upon which quantification is based. Linked-field processes are used to unambiguously link the (M+H)$^{+}$ precursor to the selected product ion. This novel MS measurement procedure can reach the low ppb level to measure ME and LE. ^{18}O-incorporated ME and LE serve as internal standards for quantification of endogenous peptides.

REFERENCES

1   D.M. Desiderio, J.Z. Sabbatini and J.L. Stein, Adv. Mass Spectrom., 8 (1980) 1298-1305.
2   D.M. Desiderio, J.L. Stein, M.D. Cunningham and J.Z. Sabbatini, J. Chromatogr., 195 (1980) 369-377.

208

3   D.M. Desiderio, S. Yamada, J.Z. Sabbatini and F.S. Tanzer, Biomed.
     Mass Spectrom., 8 (1981) 10-12.
4   D.M. Desiderio and J.Z. Sabbatini, Biomed. Mass Spectrom., 8 (1981)
     565-568.
5   D.M. Desiderio, S. Yamada, F.S. Tanzer, J. Horton and J. Trimble, J.
     Chromatogr., 217 (1981) 437-452.
6   D.M. Desiderio and S. Yamada, J. Chromatogr., 239 (1982) 87-95.
7   D.M. Desiderio and M. Kai, Biomed. Mass Spectrom., 10 (1983) 471-479.
8   D.M. Desiderio and M. Kai, Int. J. Mass Spectrom. Ion Phys., 48 (1983)
     261-264.
9   D.M. Desiderio and M. Kai, Adv. Mass Spectrom., 9 (1984).
10   D.M. Desiderio and S. Yamada, Biomed. Mass Spectrom., 10 (1983)
     358-362.
11   D.M. Desiderio and I. Katakuse, Biomed. Mass Spectrom., submitted.
12   D.M. Desiderio, I. Katakuse and M. Kai, Biomed. Mass Spectrom., 10
     (1983) 426-429.
13   D.M. Desiderio, in W.S. Hancock (Editor), Handbook of Use of HPLC for
     the Separation of Amino Acids, Peptides, and Proteins, CRC Press,
     Boca Raton, FL, in press.
14   D.M. Desiderio, J. Laughter, M.Kai and J. Trimble, J. Comp. Enchanc.
     Spectros., in press.
15   D.M. Desiderio, in J.C. Giddings, E. Grushka, J. Cazes and P.R.
     Brown (Editors), Adv. Chromatography, Vol. 22, Marcel-Dekker N.Y.,
     (1983), 1-36.
16   F.S. Tanzer, D.M. Desiderio and S. Yamada, in D.H. Rich and E. Gross
     (Editors), Peptides: Synthesis-Structure-Function, Pierce Chem. Co.,
     Rockford, Ill., 1981, pp. 761-764.
17   S. Yamada and D.M. Desiderio, Anal. Biochem., 127 (1982) 213-221.
18   S. Yamada and D. M. Desiderio, in M.T.W. Hearn, F.E. Regnier and
     C.T. Wehr (Editors), High Performance Liquid Chromatography
      Of Proteins and Peptides, Academic Press, New York, 1982, pp.
     211-219.
19   K.L. Busch and R.G. Cooks, in Tandem Mass Spectrometry, F.W.
     McLafferty (Editors), Wiley, NY (1983). pp.11-39.
20   A.M. Lawson, D.K. Lim, W. Richmond, D.M. Samson, K.D.R. Setchell and
     A.C.S. Thomas, in Current Developments in the Clinical Applications of
     HPLC, GC and MS, A.M. Lawson, D.K. Lim and W. Richmond (Editors),
     Academic Press 1980, pp. 135-153.
21   B.J. Millard, Quantitative Mass Spectrometry, Heyden, London (1978),
     171 pp.
22   D.E. Vance, W. Krivit, and C.C. Sweeley, J. Biol. Chem., 250 (1975)
     8119-8125.
23   W.C. Pickett and R.C. Murphy, Anal. Biochem., 111 (1981) 115-121.
24   D.M. Desiderio, M. Kai, F.S. Tanzer, J. Trimble and C. Wakelyn, J.
     Chromatogr., in press.
25   D.M. Desiderio, H. Onishi, H. Takeshita, F.S. Tanzer, C. Wakelyn, J.
     Walker, Jr. and G. Fridland, J. Neurochem., submitted.
26   F.S. Tanzer, D.M. Desiderio, C. Wakelyn and J. Walker, Jr., J. Dent.
     Res., submitted.
27   T. Kudo, H.-L. Chang, S. Maeda, Y. Uchida, J. Kakamae and R.
     Inoke, Life Sci., 33 (1983) 677-680.
28   T. Kudo, S. Maeda, J. Nakamae, H.-L. Chang and R. Inoke, Life Sci., 33
     (1983) 681-684.
29   H.H. Jonsson, B.S. Middleditch and D.M. Desiderio, Science, 187, (1975)
     1093 -1094.
30   D.L. Perry and D.M. Desiderio, Prostaglandins, 14 (1977) 745-752.
31   H.T. Jonsson, B.S. Middleditch, M.A. Schexnayder and D.M. Desiderio,
     J. Lip. Res., 17 (1976) 1-6.

32   M.F. Delaney, L.C. Magaz., 2 (1984) 85-86.
33   I. Lindberg and J.L. Dahl, J. Neurochem., 36 (1981) 506-512.
34   A. Dupont, J. Lepine, P. Langelier, Y. Merand, D. Rouleau, H.
      Vaudry, C. Gros and N. Barden, Regulatory Pept., 1 (1980) 43-52.
35   M.J. Kubeck and J.F. Wilber, Neurosci. Lett., 18 (1980) 155-161.
36   P.C. Emson, A. Arregui, V. Clement-Jones, B.E.B. Sandberg and
      M. Rossor, Brain Res., 199 (1980) 147-159.
37   E. Peralta, H.-Y.T. Yang, J. Hong and E. Costa, J. Chromatogr., 190
      (1980) 43-51.
38   D.B. Carr, Lancet, February 14 (1981), p. 390.
39   H.E. May, F.S. Tanzer, G.H. Fridland, C. Wakelyn, and D.M.
      Desiderio, J. Liq. Chromatogr., 5 (1982) 2135-2154.
40   M. Barber, R.S. Bordoli, G.J. Elliott and N.J. Horoch, Biochem.
      Biophys. Res. Commun., 110 (1983) 753-757.
41   D.M. Desiderio and I. Katakuse, Anal. Biochem., 129 (1983) 425-429.

Chapter 8
INSTRUMENTAL DEVELOPMENTS

8.1 INTRODUCTION

The writing of this book has been undertaken because the three different scientific fields of MS, RP-HPLC, and neuropeptide research have each been experiencing a burst of increased productive research, which in turn has resulted in increases in the detection sensitivity of analytical measurements, understanding of many neurochemical processes, and very importantly, molecular specificity. Few other areas, except perhaps the general field of arachidonic acid metabolism, are experiencing such a high level of productive research effort. On one hand, as mentioned, this dynamism is one of the main reasons for undertaking the writing of this volume, which is done as an attempt to understand and catalog the variety of the current research activity. On the other hand, this high level of activity makes it rather difficult to conclude this book because improvements are continuously appearing on the horizon. This chapter will briefly discuss several instrumental improvements which will soon impact significantly on neuropeptide analysis.

8.2 CONVERSION DYNODE

Historically, positive ions were generally measured with a mass spectrometer. More recently, negative ions have also been studied. An instrumental modification is available to effect higher sensitivity by using a conversion dynode (1).

The conversion dynode available on new commercial instruments can be retrofitted to older MS instruments. In this conversion process, positive ions strike a metal plate/conversion dynode at a voltage sufficient to produce fragment ions. Negative ions, on the other hand, will produce positive ions. In either case, a higher sensitivity is generally observed utilizing this conversion dynode. The capability may extend MS measurements of endogenous peptides discussed in Chapter 7 to the pg (fmol) level.

8.3 HIGHER MASS CAPABILITIES

The development of commercially available MS instrumentation is constantly incorporating improvements in magnetic, electronic, and machining technology.

It is now possible to obtain single-charged ion masses up to 10,000 mass units at full accelerating voltage, mainly because of the significant improvements in magnet design (see Chapter 6). This mass range is significant for neuropeptide research because it now permits more extended peptide studies to include compounds such as precursors which lead to biologically significant peptides such as opioid peptides, substance P, and insulin.

Laminated magnets are used to effectively reduce eddy currents and fringing magnetic fields, and to increase scan speeds up to 100 msec decade^{-1} in an effort to improve performance specifications compared to solid core magnets. Appropriately-shaped magnet pole pieces are used to reduce fringe fields and to increase mass range. Non-normal ion entry angles plus inhomogeneous magnetic fields are used to achieve an increased mass range. Computer acquisition at digitization rates near 200 megahertz maintain mass-measuring accuracy at these previously unattainable high mass ranges and scan speeds.

All of these instrumental improvements have recently become commercially available, and it can be expected that significant improvements will continue in our abilities to measure endogenous peptides with these improved instruments.

8.4 HIGHER SENSITIVITY

Use of the conversion dynode and microcomputer processing will lead to increases in the sensitivity of our peptide measurements. Picogram (femtomole) amounts of the neuropeptides will be soon quantified with retention of that optimal molecular specificity which is offered by monitoring an amino acid sequence-determining ion in the CAD-linked field-SIM mode. It is not inconceivable to expect that this type of MS measurement mode will soon be equivalent in sensitivity to RRA and RIA methodologies, but with the significant advantage of virtually unambiguous molecular specificity. As discussed above (Section 7.8), when all pertinent operational factors of the RIA and MS measurement methods are considered, the cost of analysis per unit sample of the two methods is equivalent.

8.5 NEGATIVE IONS

Generally, all of the ionization methods used in MS will produce both positive and negative ions. The two FAB-produced ion modes for peptides have been studied (2, 3) and are available to the researcher to provide yet another dimension to the type of molecular specificity that may be obtained with the off-line or on-line combination of HPLC and MS. In some cases, ions are formed in the negative but not the positive ion mode, some in the positive but not the negative, and some ions in both modes. Judiciously selected negative ions may be used to further increase the molecular specificity and detection sensitivity which is available for MS measurement of endogenous peptides.

8.6 THE ON-LINE COMBINATION OF LIQUID CHROMATOGRAPHY AND MASS SPECTROMETRY

8.6.1 Introduction

At this point in the development of the theme of this book, this is a natural place and format to discuss one instrumental development which will have a significant impact on the MS analysis of endogenous neuropeptides. The on-line combination of HPLC-MS is a natural instrumental combination that effectively couples the separation power of HPLC with a specific detector (MS) that is capable of achieving high levels of sensitivity and molecular specificity (4). Up to now, the principles of an off-line combination of HPLC and MS have been discussed. The driving force to provide an on-line combination of HPLC and MS is the very desirable decrease in the number of experimental steps and manual interventions which are required for manipulation of small amounts of precious endogenous peptides. In any event, the coupling of these two instrumental methods is quite difficult to achieve and indeed, it is the author's opinion that the rate of the development of achieving an efficient on-line coupling of the two instrumental techniques is relatively slow as compared to the corresponding development of the combination of GC with MS (5). Nonetheless, investigators are creatively approaching this problem and it is simply a matter of time until this combination technique will be in the hands of many investigators. The literature pertaining to on-line LC-MS has been reviewed recently (4, 6-8).

8.6.2 Basic considerations

In principle, there are two different methods to achieve the on-line combination of HPLC with MS. On one hand, a direct liquid introduction (DLI) interface performs exactly what the name implies, that is, the liquid eluting from the HPLC column is introduced directly into the MS instrument. The solvent is removed by some appropriate process such as nebulization, thermospray, etc. On the other hand, the effluent from the HPLC column may be deposited upon a moving belt which transports a chromatographic fraction from the outside high pressure region of the HPLC unit through an interface into the high vacuum portion of the MS where ionization is effected. These two separate techniques, DLI and moving belt interfaces, will be discussed in greater detail in the next section.

There are five other operational parameters which must be thoughtfully analyzed and experimentally optimized for any particular type of on-line HPLC-MS development and experimentation. These parameters include a normal HPLC column versus a microbore column (9, 10); the use of either a high voltage magnetic MS instrument or a low voltage quadrupole mass spectrometer; use of either the split or splitless mode for the interface following the HPLC column; the use of either volatile or nonvolatile buffers for the HPLC separations; and the selection of an ionization process, which plays a significant role in that either EI or CI may be used to ionize the HPLC fraction as opposed to a surface method such as SIMS or FAB MS.

8.6.3 Direct liquid introduction interface

One of the sets of experiments to study the DLI of liquid solutions into a mass spectrometer involved the use of CI MS (11-15). The CI source used as a coupling for LC-MS followed the work with LC-MS in the EI mode. CI LC-MS was developed to such a point whereby polypeptide sequencing was possible. For determining the amino acid sequence of a peptide, the authors felt that the information from a CI mass spectrum appeared to be at least as valuable as that information which is derived from EI. DLI has the further advantage that the resulting mass spectrum allows sequencing of less volatile samples such as underivatized pentapeptides. In this study, the zwitterionic character of the peptide was eliminated by acetylation of the free amino group and esterification of all the free carboxyl groups. The HPLC solvents include acetonitrile and water. A capillary splitter interface was used in front of the HPLC UV detector to introduce five to ten microliters per minute of the LC effluent continuously into a CI mass spectrometer. Because only about one or two percent of the peptide sample enters the mass spectrometer, the sensitivity of the method is decreased by a factor of 100. Mass spectral scans were cycled continuously during an LC run, and two or three corresponding mass spectra are obtained per HPLC peak. The HPLC solvent serves readily as the CI reagent, although other reagents can be added concurrently to alter that type of ionization.

The critical instrumental parameters necessary for combining HPLC with a mass spectrometer have been optimized. In one case, discussion centered on whether or not the LC solvent should be removed in the interface (16). It is stated that the basic philosophy in the development of the first LC-MS systems included:

(i)   introduction of the total HPLC column effluent into a separator;

(ii)   selected removal of the chromatographic carrier;

(iii)   vaporization of the neutral analyte;

(iv)   ionization and mass analysis.

These initial developments led to transport systems which are based on the use of a moving belt or wire for conveying the solution, then the solute through these four stages. In the cases of ionization, it was soon realized that pre-formed ions already existing in the solution could be directly vaporized and analyzed when enough energy is supplied to the liquid solution. Several methods are available for ionization and include electrospraying, electrohydrodynamic ionization, FD, thermospray ionization, and FAB MS.

When interfacing a chromatographic column to a mass spectrometer, it is important to realize that the chromatographic column performs both as a separation and as a dilution unit (17). The problems involved with interfacing an HPLC unit to a mass spectrometer stem from the relative incompatibility between an effluent solution eluting from the HPLC chromatographic column and the low pressure gas inside the source of the mass spectrometer (17). These two features make the coupling of the two instruments even more difficult as compared to interfacing a GC to an MS.

It is important to consider the main constraints which are introduced by the chromatographic process as well as to couple those constraints with the minimum requirements that the interface must possess in order to accurately describe the degree of competitiveness that this interface would possess and to suggest compromises acceptable from the chromatographic point of view.

When combining an HPLC unit with an MS, it is important to realize that solvent expansion occurs spontaneously (irreversibly) from a thermodynamic point of view. The decrease in entropy which arises from the separation is more than compensated for by the increase in entropy deriving from the dilution in the mobile phase. Whatever interface is selected for the combined LC-MS methodology, the use of a non-volatile buffer seems prone to generate considerable troubles. Furthermore, although gradient elution is much talked about in HPLC research circles, that form of separation is relatively rarely used. The detection limit of a chromatographic detector is defined as that mass of a compound that generates a signal which is twice that of the noise.

The chromatographer is always surprised by the low ionization yield of the mass spectrometer (E), the number of ions collected on the mass spectrometer detector per number of molecules which is introduced into the ion source. It is important to review briefly the various sources of those losses (17):

(i)   To detect a signal at a given mass, and to calculate the location of the signal maximum corresponding to the molecular weight, one needs approximately 100 ions at the detector entrance slit; practically all ions reaching this slit are detected.

(ii) The object and image slits in most mass spectrometers are of a rectangular cross section. When a mass spectrometer is scanned, the product of these two (presumably identical) slits is a triangle. It is a requirement to have 200 ions enter the analyzer, where losses in the analyzer itself are assumed to be negligible.

(iii) The extraction yield of ions from the source to the analyzer across ion optics (ion-focusing) is about 10%. Therefore, 10 x 200, or 2,000 ions are needed in the source.

(iv) For identification purposes, a complete scanned mass spectrum is needed and the previous figure (2,000 ions) is applied to those ions which account only for small peaks in the mass spectrum. Peaks that are 10% of the base peak must be detectable. Therefore at least 2×10^4 molecular ions must be formed during the time when the corresponding mass is scanned.

(v) The ionization yields vary widely with both the ionization method used and the particular compound being analyzed. While that efficiency may be nearly one for electron capture of haloaromatics, it may on the other hand be as low as 10^{-4} for the EI of many other compounds. If these two values are averaged to 10^{-3}, then 2×10^7 molecules should be present during the scan.

(vi) The time to scan one mass in a spectrum is about one msec, and the introduction of sample molecules into the source must therefore proceed at a rate of approximately 2×10^{10} molecules sec^{-1}.

The maximum concentration C_m of the Gaussian band of a solute of retention volume V_r and efficiency N is:

$$C_M = m \ N^{\frac{1}{2}} / \ V_r \cdot (2\pi)^{\frac{1}{2}} \tag{1}$$

where m is the sample mass. If the column capacity is k', and the liquid cross-section of the column is s, then:

$$C_M = (m/s) \ (N^{\frac{1}{2}}) \ [L \ (1 + k') \cdot (2\pi)^{-\frac{1}{2}}] \tag{2}$$

where L and u are the column length and the solvent velocity, respectively. The mass flow-rate of sample into the mass spectrometer source is then the product $C_M \ F$, where F (= Su) is the solvent flow-rate. With a splitting ratio r, the mass flow-rate of sample into the source is:

$$dm \ / \ dt = C_M Fr = (m \cdot u \cdot r \cdot N^{\frac{1}{2}}) / \ [L \cdot (1 + k') \cdot (2\pi)^{\frac{1}{2}}] \tag{3}$$

Comparing equation three with the condition (vi) above;

$$N' \cdot (m \cdot u \cdot r \cdot N^{\frac{1}{2}}) / [M \cdot L \cdot (1 + k') \cdot (2\pi)^{\frac{1}{2}}] = 2 \times 10^{10} \qquad (4)$$

where M is the molecular weight of the solute and N' is Avogadro's number.
With L = 15 cm, N = 1.5 x 10^4 plates, u = 0.05 cm per second, r = 1, M = 500,
and k' = 1, then m = 2 x 10^{-10} grams = 200 pg. This level of sensitivity is
in general agreement with the specifications of modern mass spectrometer
instruments which give the detection limit of 100 pg of methyl stearate (M
= 298, m = 120 pg), although the latter specifications may not have been
calculated with the rather favorable chromatographic conditions which are
selected above: narrow peaks with small retention to give large maximum
concentration.

On the other hand, if, as we have discussed in previous chapters, the
mass spectrometer operates not in the scanning mode, but rather as a true
selected ion monitor, then a smaller amount of sample is necessary: with a
one second time-constant, 1000 times less, or around 100 fg to 1 pg would
suffice for detection. The detection values are similar for either a magnetic
instrument or for a quadrupole MS. The only possibility of improving these
detection limits in a significant fashion is whenever a very efficient
ionization technique is utilized; a fact that explains why haloaromatics
can be determined at the fmol level with negative ions.

A high speed DLI LC-MS interface is described (18). At the exit of
the HPLC probe, a heated pre-evaporation chamber is provided where the
molecules have a low speed in a high pressure region. Following that region
is a short transition zone from which the molecules enter a high speed - low
pressure region before entering the ion source. This latter region is the
controlled desolvation chamber. The maximum velocity in the transition zone
is sonic velocity (Fig. 8.1)

8.6.3.1 Controlled desolvation chamber

A new desolvation chamber for droplet focusing or Townsend discharge
ionization is described as an interface for DLI into a mass spectrometer.
This new desolvation chamber contains a conically-shaped volume and derives
from considerations that utilize a slightly different concept because
preliminary experiments show that the liquid droplets are highly
positively-charged during nebulization and that physical contacts with the
walls of the desolvation chamber should in general be avoided. A positive
potential on the walls therefore can be used to repel the droplets. On the
other hand, by adoption of the appropriate conical geometry, droplets could

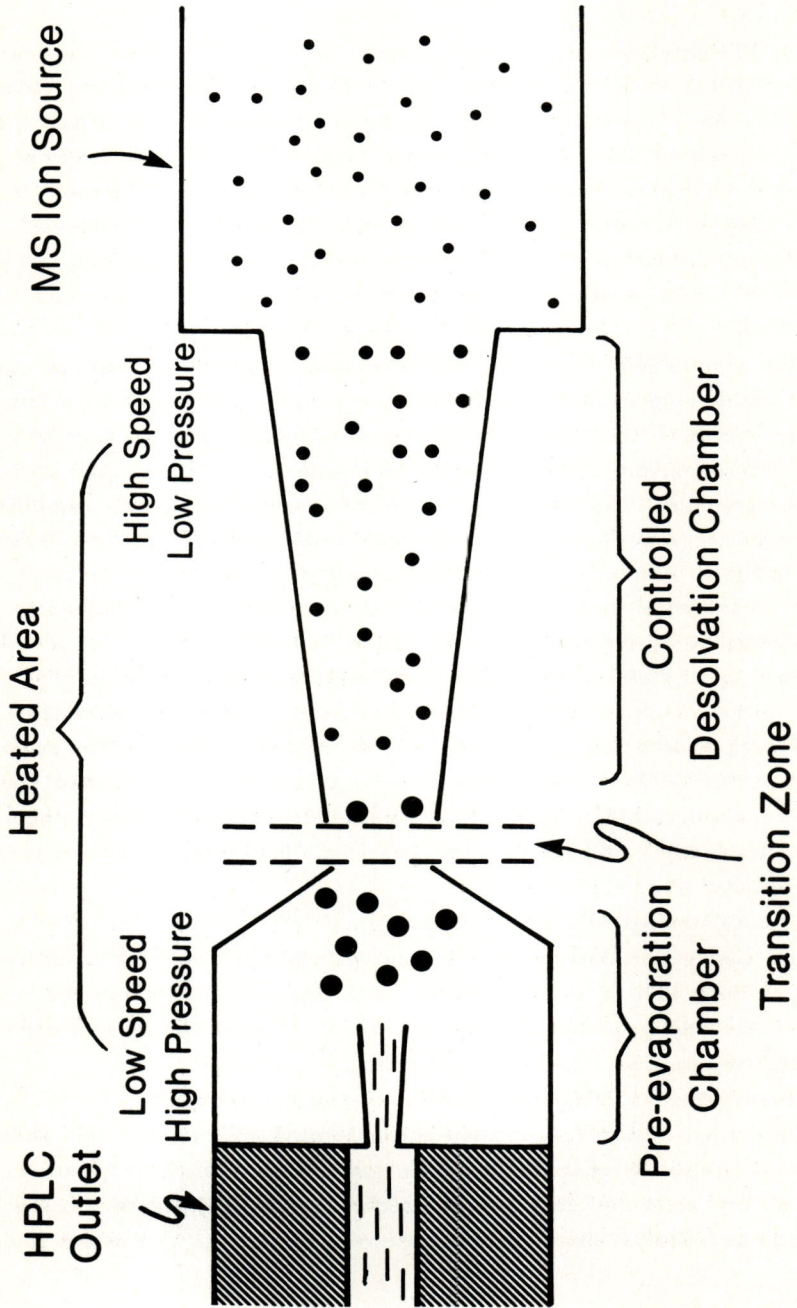

Figure 8.1. Schematic of high speed DI HPLC-MS interface (18).

either be focused along the drift tube axis or rather, converge to an appropriate location.

Other considerations of LC-MS interfacing have been published by this research group (19-21).

8.6.3.2 Thermospray

Thermospray is defined as the complete or partical vaporization of a liquid stream by heating it as it flows through a capillary (22). Thermospray has recently been demonstrated to be versatile interface for combined LC-MS, where the heat is supplied electrically and controlled by an electronic feedback system to maintain a constant level of vaporization. The optimal temperature for thermospray is a function of the solvent composition, flow rate, capillary dimensions, and the sample to be analyzed. Thermospray interfaces were first heated by lasers, then by oxy-hydrogen torches, and most recently by cartridge heaters.

One of the first papers in this work describes a new soft ionization technique for the MS of complex molecules (23). Briefly, the effluent from an HPLC column enters the vaporizer via a steel capillary tube which is partially immersed in a copper cylinder that is heated to about 1000° C by four small oxy-hydrogen flames. As a result of rapid heating, a jet of vapor and aerosol is produced near the exit of the stainless steel tube. The jet is further heated as it passes through the channel in the copper. It then undergoes an adiabatic expansion, and a portion passes through a skimmer to the ion source, where the beam impinges on a nickel-plated copper probe which is electrically heated to about 250°C. While the ion source of the mass spectrometer is equipped with an electron gun for producing ions for normal CI operation, in the work described in this manuscript, it is turned off. When the charged particle undergoes a high energy impact with the heated probe, it is wholly or partially vaporized and some of the resulting molecules are ionized. Spectra are obtained for dinucleosides (CpG; ApU) and the pentapeptide LE. Background ionization of the solvent is low, and this fact, coupled with reduced amount of fragmentation that is generally observed, lead to the surprising result that the detection limit for the $(M+H)^+$ of most of the nonvolatile substances investigated is substantially lower with the electron beam turned off!

Both positive and negative CI mass spectra were obtained for nonvolatile biologic samples in the one to ten nanogram range for full mass scans and subnanogram quantities with SIM (23). It is noted that vaporization is possible without pyrolysis because the sample spends a very short time in the high temperature region in the thermospray interface. During that transit time, these samples are protected from overheating by the solvent. The liquid entering the hot region is heated from ambient temperature to the vaporization point in a few msec, and the vapors expelled from the heat region a msec or less after they form. Experiments show that the detection limit (signal-to-noise ratio of two) for arginine is 500 pg and for methionine, 2.5 ng. If the SIM measurement mode would be used for only the $(M+H)^+$ ion intensity, the detection limits are predicted to be more sensitive by about one order of magnitude. In this study, phosphate buffers cannot be used, but the volatile buffers ammonium formate and ammonium acetate are used extensively with essentially no difficulties. The major limitation to the operation of the mass spectrometer is that the solvent vapor is also the CI reagent. This dual role of the same substance can cause difficulty in the simultaneous optimization of the HPLC separation and the MS detection.

As mentioned above, cartridge heaters are used to replace the oxy-hydrogen torches in a thermospray interface which is readily adapted to quadrupole mass spectrometers (24). HPLC effluents are thermosprayed directly into the ion source and the excess vapor is pumped away by an added mechanical pump which is directly coupled to the ion source through a port opposite the electrically-heated themospray vaporizer. When used with mobile phases containing a significant concentration of ions in solution (ca. 10^{-4} to 1M), no external ionizing source is required to achieve detection of many nonvolatile solutes at the subnanogram level. The mass spectrum of the tridecapeptide renin substrate is shown and represents the largest molecule which has been successfully detected by using thermospray ionization. The $(M+H)^+$ weight of the renin substrate is 1,758 mass units.

A review of all the ionization techniques currently available for MS for nonvolatile molecules has been published (25). A common feature of all of the ionization methods appears to be the direct production of ions from a condensed phase without formation of a neutral gas-phase molecule as an intermediate. An attempt is made to present a unified ionization model which at least qualitatively accounts for the results obtained by the various techniques. A laser desorption mass spectrometer is interfaced to an HPLC unit using a moving stainless steel belt, where samples are sprayed online onto the belt under partial vaccum through a thermospray vaporizer. Laser desorption occurred with a laser. Two modes of operation are presented for the thermosprayer, one with (closed) and one without (open) the transfer line. Using an electrospray method for sample deposition, comparisons were made amongst the electrospray, closed, and open modes. While it was assumed that the electrospray method has a 100% efficiency to transfer the sample, the closed mode correlates to 40% and the open mode to 20% efficiency of covering the belt. A major limitation of the present moving belt system was found to be that the HPLC separations must be accomplished in less than nine minutes in order to prevent sample components from overlapping each other on that length of belt.

Triply- and quadruply-charged molecular ions of nonderivatized glucagon (molecular weight equals 3,483 a.m.u.) are observed (26). In another experiment, a procedure is presented for peptide sequencing by utilizing an immobilized exopeptidase column which is directly coupled to a thermospray mass spectrometer (26). Amino acid sequence determination starting at the C-terminus was effected in this manner. This experimentation parallels that found previously for an immobilized enzyme and which is discussed in Section 7.10 (27). The solutions of underivatized peptides are injected through a column containing immobilized carboxypeptidase Y and the amino acids were released, starting from the C-terminus of the peptide chain, and directly transported by a continuously flowing aqueous buffer into a thermospray mass spectrometer where the $(M+H)^+$ ion of each amino acid is determined. Temperature of enzymolysis is a factor because, at 22°, only two amino acids are found whereas at 42°, five amino acids are determined from angiotensin.

8.6.3.3 Microbore LC-MS

The effluent from a microbore HPLC which operates at a flow rate of eight microliters min^{-1}, is introduced into a quadrupole mass spectrometer ion source operating in the CI mode (28). This paper studied evaporation of liquids from a capillary into a vacuum, where a jet was observed inside a glass envelope. The jet appears to travel straight in air. However, when a vacuum is applied, that jet starts to bend. This phenomenon is explained by considering the fact that both large and small droplets are formed and that the small droplets evaporate more rapidly than the larger droplets. If the small droplets are predominantly formed on one side of the jet, owing to the irregular shape of the orifice, the evaporating vapors from the small vapors push the large droplets away from the axis of the jet when a vacuum is applied.

Instrumental and analytical advantages are taken of microbore HPLC coupled through a DLI interface to a CI mass spectrometer (29). The analytical capabilities of a micro-HPLC which is interfaced to an unchanged quadrupole mass spectrometer demonstrate continuous monitoring of the total micro-HPLC effluent. Full scan CI mass spectra of drugs are obtained in the range of one to five nanograms. A microbore HPLC flow rate of eight microliters per minute is utilized and SIM provides a 20 picogram detection limit of a tranquilizer.

Conventional HPLC instruments generally operate at flow rates of 0.5 to two milliliters per minute. It is not practical to introduce the total HPLC effluent through a DLI interface. However, microbore packed HPLC columns operate at eluant flow rates of two to forty microliters per minute and are ideal for DLI. Because the entire effluent from a micro-HPLC can be directed into the ion source for mass spectral analysis, a 100-fold increase in detection sensitivity may be realized (30). Conventional HPLC systems yield peak volumes of 0.5-2.0 ml, where that range depends upon the flow rate and column efficiency. On the other hand, utilizing microbore HPLC, peak volumes are found in the range of 10 to 30 microliters. The smaller volume in the latter case provides a much more concentrated solution passing through the detector with a concomitant increase in the detection limit (31).

8.6.3.4 Liquid chromatography-mass spectrometry-mass spectrometry

In another study determining sulfa drugs in biologic fluids, it is noted that one limitation results from the DLI technique of LC-MS because the mass spectra are very simple and they lack sufficient specificity for elucidation (32). Furthermore, it is always possible that co-eluting compounds will appear superimposed in the mass spectrum, a process which renders the interpretation rather difficult. The novel technique of combined liquid chromatography-mass spectrometry-mass spectrometry (LC-MS-MS) offers the analytical capability of also providing CAD mass spectra of the characteristic LC-MS ions for each of the components of interest (see Chapter 6). In this study, the potential of a more open atmospheric pressure ionization (API) source design is utilized in combination with HPLC and MS in a TSQ instrument. Nitrogen is used as a collision gas with an effective target thickness of approximately 2×10^{14} molecules centimeter^{-2}. This study is prompted by the extreme paucity of fragmentation that occurs in the positive CI LC-MS mode.

8.6.3.5 Liquid chromatography-negative chemical ionization mass spectrometry

In another study utilizing HPLC, negative ion CI MS is undertaken because of the need to analyze mixtures of explosives (33). Post-explosion analysis of residues is very relevant in criminalistic bombing. A simplified construction of a new micro LC-MS probe is presented, where the volume of the HPLC microcell is calculated to be 1.2 microliter, and which maintains the one centimeter path length of the conventional cell, where the latter has a 15 microliter volume (34).

A desolvation chamber of new design sprays the liquid chromatographic effluent into droplets through a pin-hole diaphragm (35). One of the most effective ways of controlling the desolvation of the solute droplets is to use a heated zone, usually called a desolvation chamber, in which the droplets acquire a high speed before entering the CI source. Three different desolvation chambers were studied and include the standard Hewlett-Packard, extended, and solvent-stripping models. A review of the micro LC/MS methodology used in drug analysis and metabolism studies has been published (36) and the quantitative analysis of betamethasone in equine plasma and urine by DLI micro-LC/MS is presented (37).

A new version of the thermospray LC-MS interface is described, where this version differs from the previously described interfaces in that it is a dual- purpose probe-type interface which is introduced conveniently into the mass spectrometer via a standard direct insertion probe inlet (38). The device is a dual-purpose LC-MS interface and it can provide conventional DLI LC-MS, or the copper vaporizer may be heated electrically to produce thermospray ionization. Detection limits are currently about 100 nanograms in the thermospray mode for nonvolatile, labile compounds using two mm i.d. microbore LC columns and a flow rate of about 150 microliters per minute.

8.6.3.6 Ultrasonic interface

An ultrasonic spraying device was constructed to overcome difficulties in spraying aqueous solvents into the vacuum system of a mass spectrometer. The ultrasonic vibration is achieved by means of a magnetoconstriction in the nickel inlet tube (39).

8.6.3.7 Segmented wire interface

It has been noted that the DLI technique yields an intrinsically simple implementation and that the moving belt interface has significant advantages in several parameters including solute concentration - a process that introduces a greater proportion of the sample into the mass spectrometer ion source (39, 40). This group approaches the interface problem by preconcentrating a liquid stream and introducing the effluent into the MS by means of DLI. The viewpoint is that the advantages of both ordinary DLI and the moving belt technique would be combined. The described interface device concentrates a liquid stream by allowing it to flow down a resistance-heated stationary wire having three successively-decreasing diameters of 0.8 mm., 0.6 mm., and lastly, 0.3 mm., where each section has a length of 15, 7.5, and 1.5 cm., respectively. When a drop becomes too large, it flows either down the wire or along the outer surface of the DLI probe and is lost. A light-emitting diode and photocell are fitted at the gap and sense the size of the drop. Electronic feedback from the photocell controls the current through the wire to hold the drop size constant. The instrumental parameters are met only with the wire that was not of homogeneous construction and the three-segment wire design was arrived at empirically. A description is provided where the performance of the concentrator wire is tested in an effort to determine the maximum flow which could be accomodated and still maintain a 95% evaporation of the solvent, corresponding to a 230-fold increase in sample concentration.

8.6.3.8 Nebulization

The design of a DLI interface is described which uses a jet of helium gas to aid in the nebulization of the vaporizing HPLC effluent and sample into the MS source (41). Interfaces used in coupling HPLC with MS fall into two basic categories: transport interfaces and DLI interfaces (DLI). A number of systems have been used in the DLI approach for introducing the LC effluent into the mass spectrometer source and include formation of liquid jets through either viscous flow capillary or 1.5 micron diaphragms and vacuum nebulization techniques.

8.6.3.9 Supercritical fluid injection

Direct fluid injection interfaces operating under supercritical conditions have been designed for use with MS (42). Direct fluid injection (DFI) mass spectrometry utilizes supercritical fluids for solvation and transfer of materials to a CI MS source. Supercritical carbon dioxide with isobutane as the CI reagent gas is used and DFI MS/MS is illustrated for major ions in the isobutane CI spectrum of T2 toxin. More polar compounds may be analyzed using supercritical ammonia. This alternative HPLC-MS approach uses a supercritical fluid or "dense gas" for efficient transfer of material to the gas phase in the CI source. The DFI method allows mass spectra to be obtained for essentially any compound which is soluble in the supercritical fluid and hence allows a rapid qualitative evaluation of fluid phase solubility.

At high pressures and above the critical temperature, the resulting fluid or dense gas attains a density approaching that of a liquid, which has relatively strong internal molecular interactions and therefore assumes some of the properties of a liquid. In describing the DFI process, the supercritical fluid exits from a 50 to 500 atmosphere pressure region, through a hole into a region in which a Mach disk is produced where initial clusters have a diameter of approximately 30 angstroms. The shock waves resulting after the Mach disk then transfer into a molecular spray which enters a CI region of a mass spectrometer at a pressure of approximately one torr. The Mach disk is characterized by two phenomena - the destruction of the highly-directed jet and the collisional energy transfer resulting in redistribution of the directed kinetic energy of the jet among the various translational, vibrational, and rotational modes of the molecule.

The use of capillary column supercritical fluid chromatography/mass spectrometry can obviate the difficulties associated with previous interfaces and allows a simple interface readily adapted to existing GC/MS systems (43). The combination of SFC with mass spectrometry offers the following potential advantages relative to GC/MS or LC/MS methods:

a. high molecular weight, polymeric, polyfunctional, and thermally labile compounds can be separated, as well as the more volatile species;

b. capillary SFC columns can provide greatly enhanced chromatographic efficiency relative to HPLC due to solute diffusivities which are about 100 times greater in the supercritical fluid than in the corresponding liquid phase and similar to the gas phase;

c. soluting power of the mobile phase can be readily controlled with pressure programming. Mixed mobile phases, gradient, and temperature programming are also feasible;

d. SFC using a capillary column provides low mobile phase flow rates which, when coupled with high mobile phase volatility, allows optimal interfacing of SFC and MS.

Optimal mobile phase flow rates of five to 80 microliters per minute (supercritical fluid), depending on column diameter and pressure, may be obtained in this combination instrument. The SFC mobile phases used in this work are isobutane and normal pentane. Initial evaluations of the capillary SFC-MS interface demonstrate that it is mechanically simple and reliable. Polyaromatic hydrocarbon studies indicate that a detection limit of this system is approximately one picogram is achieved.

8.6.3.10 Examples of LC-MS analyses of compound types

Glucuronides are characterized using a thermospray LC-MS interface (44). Ten nanograms of the glucuronide injected onto a column suffice for characterization by a scanned mass spectrum. Unlike many of the other currently available LC-MS interfaces, the thermospray ionization interface allows 100% of an aqueous effluent to enter the mass spectrometer at flow rates up to 1.5 ml per minute.

Nanogram amounts of the peptides LE, ME, and alpha-amanitin are obtained by direct LC-MS (45). Both $(M+H)^{+}$ and $(M-H)^{-}$ ions are obtained in the positive and negative ion modes, respectively.

8.6.4 Moving belt interfaces

8.6.4.1 Introduction

The addition of a modified segmented-flow extractor between an HPLC and a MS permits the direct coupling of an HPLC operated in the RP mode to a mass spectrometer, without compromising the operational characteristics of either instrument (46). Ion-pairing techniques were studied and demonstrate compatibility with on-line LC–MS (47).

Several approaches had been explored for the direct LC–MS combination:

(i)   A moving wire/belt system that transports the LC effluent through a series of vacuum locks for the solvent to be evaporated. The sample is then introduced into the ion source for analysis by either EI or CI.

(ii)   A 1% split of the LC effluent into the mass spectrometer where the LC solvent is used as a CI reagent gas.

(iii)   A vacuum nebulizing interface to introduce total effluent from a microbore HPLC into the ion source.

(iv)   The system that converts the LC effluent into a molecular beam by forcing it through a nozzle restriction, followed by flash evaporation using a laser beam or sonic radiation, and then MS analysis in either the EI or CI mode.

(v)   Direct evaporation of the total effluent into the mass spectrometer ion source, followed by CI under API conditions.

(vi)   A silicone membrane enrichment device that removes the solvent and permits preferential entry of the solvent into the mass spectrometer.

The segmented-flow LC–MS interface desalts the organic phase with high efficiency. This fact is demonstrated by analyzing sodium ions and phosphate ions at the 0.3 ppm and 10 ppm levels, respectively.

In a further development of this type of LC–MS interface, a specially designed nebulizer is constructed for deposition of the effluent from the HPLC column onto a moving belt (48). Another aerosol liquid deposition device was described (49). Measurement of the solvent transfer efficiency to the belt is performed by first spraying a peak from the HPLC column onto the moving belt, and then comparing that peak area with the area obtained when depositing the same mass of sample onto the belt with a syringe, where the syringe method is assumed to provide 100% efficiency in the transfer of the solute onto the belt. A belt interface can be thought of as consisting of four basic steps, each of which may potentially affect chromatographic performance:

(i) transfer or deposition step;

(ii) evaporation of the solvent remaining on the belt with an infrared heater;

(iii) passing of the solvent into the vacuum lock system;

(iv) desorption of the sample from the belt into the ion source.

The influence that the deposition step can have on the performance of the LC–MS interface is critical, when comparing the conventional method of flowing the effluent onto the belt in a continuous stream versus spray deposition. Dispersion of the LC effluent into a fine mist can provide an efficient evaporation step. It is found that the dimensions of the orifice through which the liquid flows must be minimized to prevent liquid from accumulating on the glass tip, and to allow formation of the smallest possible droplets. Droplet formation on the belt behind the tip is minimized by using a 60° angle between the spray tip and belt. Attention to these critical experimental steps is of primary importance in order to obtain good chromatographic fidelity including peak shape, variance, area, and reproducibility.

8.6.4.2 Peptide studies

Using a quadrupole mass spectrometer which is outfitted with a moving belt LC-MS interface, N-acetyl-N,O,S-permethylated oligopeptides were analyzed (50, 51). Isobutane CI yields good intensity of $(M+H)^+$ and $\underline{N}$- and $\underline{C}$-terminal ions. In addition, C-methylated peptides are separated by LC.

Permethylated peptides and peptide mixtures have been studied employing normal phase chromatography and a moving belt LC-MS interface (52). Eighteen different peptides, ranging in size from di- to heptapeptides, were studied, and it was found that on-line LC-MS ammonia CI spectra produced complete or almost complete amino acid sequence-determining information.

8.6.4.3 Analysis-of-variance of system components

Extra-column band-spreading which occurs on an HPLC-MS moving belt interface was analyzed by a numerical evaluation of the system variance (53). Spraying effects are evaluated by considering the increase of the variance (or the second moment of mass) of the chromatographic band. Variances are additive when contributions are independent.

8.6.4.4 Review of LC-MS transport devices

HPLC-MS interfaces with transport devices are reviewed (54) and a comparison is made of moving belt interfaces for LC-MS (55). In the interfacing of the HPLC to a mass spectrometry, three fundamental problems must be overcome:

(i)   how to make the mass spectrometer, which can handle 20 ml per minute of gas if configured for CI, compatible with solvent flow rates of the order of one ml per minute, which result in gas volumes in the range of 150 to 1,200 ml per minute, depending on the solvent used;

(ii)   introduction of the solute into the mass spectrometer so that mass spectral information can be obtained and the solute does not undergo thermal decomposition;

(iii)   coupling of the HPLC with the mass spectrometer so that chromatographic resolution and performance is maintained.

It is noted that a further problem which occurs is that, with aqueous solvent systems containing more than 50% water, beading of the solvent on the belt causes pressure fluctuations in the ion source resulting in poor mass spectral data. Although microbore HPLC was initially utilized with the moving belt system, it has had little subsequent use.

A table in that review (55) collects the applications and corresponding references of HPLC using transport type interfaces and includes the following compound types: aflatoxins, Amarylidaceae alkaloids, antibiotics, aromatic acids, bile acids and their conjugates, carbamate pesticides, chinchona alkaloids, chlorinated phenols in urine, coal liquefaction products, dinitrophenyl hydrazones, drugs, effluents, ergot alkaloids, glycosides, herbicides, lipids, liquid crystals, natural coumarins, nucleosides, peptides, pesticides, petroporphyrins, polychlorinated biphenyls and their metabolites, polynuclear aromatics, rotenoids, steroids, sugars, triglycerides, and waxes.

8.6.4.5 Ribbon storage device

A novel ribbon storage interface is described (56, 57). The distance between the MS and the HPLC unit approximates five feet. This interface is designed for use with SIMS as well as with conventional EI. The LC-MS interface includes a 120 cm region at atmospheric pressure, provision for aerosol deposition of the HPLC effluent to allow the complete evaporation of the LC mobile phase before the first vacuum slit, and a 320 cm total length to allow the storage of chromatographically-separated materials. Ten picograms of amino acids are detected. The long ribbon also allows temporary storage of five to 60 min. of HPLC separations on the ribbon for subsequent reanalysis by SIMS or EI.

A new method for ribbon cleaning using vapor deposition of a thin layer of silver is described; it reduces background from contaminants and residues on the ribbon and is superior to heaters or solvent baths. The LC eluent is deposited on the ribbon surface by an aerosol liquid deposition device. The heart of the interface is a ribbon (0.63 cm wide, 0.0087 cm thick, and 320 cm long) which is spot-welded to form a continuous loop. Ribbons of high-purity nickel, molybdenum, and platinum are found to have acceptable mechanical properties; most of the work is done with high-purity nickel.

8.6.4.6 LC-MS ionization methods

Ionization methods available for LC-MS are reviewed (58). Nine different methods are described: DCI, laser desorption (LD), FD, electrohydrodynamic ionizaiton, [252]Californium plasma desorption, SIMS, FAB, API, and the thermospray ionization technique.

8.6.4.7 Conclusions

At the present time, no one of the interface devices described above has taken charge of the field to become universally useful in laboratories around the world. Each one of the proponents of the different on-line LC-MS techniques can describe the advantages and disadvantages of each system. It is clear from the literature that more developmental time is needed. That hesitation of employing the commercially available HPLC-MS interfaces notwithstanding, it is quite clear that this type of interface, once appropriately developed and utilized, will significantly increase the use of MS as the detector of choice for HPLC for the measurement of endogenous neuropeptides.

8.7 SUMMARY

Several MS instrumental developments are on the horizon and include higher sensitivity and higher mass specifications. Several types of interfaces for on-line LC-MS are being developed, and experimentalists are eagerly awaiting that one interface that will emerge and exhibit excellent dependability, sensitivity, mass range, cost, and efficiency characteristics.

The basic developments in the analytical instrumentation will be translated rapidly into use for measurement of endogenous biologically important peptides. Dependable measurements of endogenous compounds will in turn hasten our understanding of many biological events and even more important, will provide a firm and dependable underpinning of molecular structures.

REFERENCES

1   D.F. Hunt, G.C. Stafford, F.W. Crow and J.W. Russell, Anal. Chem., 48 (1976) 2098-2105.
2   M. Barber, R.S. Bordoli, G.V. Garner, D.B. Gordon, R.D. Sedgwick, L.W. Tetler and A.N. Tyler, Biochem. J., 197 (1981) 401-404.
3   I. Katakuse and D.M. Desiderio, Int. J. Mass Spectrom. Ion Phys., 54 (1983) 1-15.
4   D.M. Desiderio and G.H. Fridland, J. Liq. Chromatogr., in press.
5   J.T. Watson and K. Biemann, Anal. Chem., 36 (1964) 1134-1137.
6   C.G. Edmonds, J.A. McCloskey and V.A. Edmonds, Biomed. Mass Spectrom., 10 (1983) 237-252.
7   P.J. Arpino, Trends Anal. Chem., 7 (1982) 154-158.
8   W.H. McFadden, J. Chromatogr. Sci., 19 (1980) 97-102.
9   C. Eckers, K.K. Cuddy and J.D. Henion, J. Liq. Chromatogr., 6 (1983) 2383-2409.
10   C. Eckers, D.S. Skrabalak and J.D. Henion, Clin. Chem., 28 (1982) 1882-1886.
11   B.G. Dawkins, P.J. Arpino and F.W. McLafferty, Biomed. Mass Spectrom., 5 (1978) 1-6.

12 M.A. Baldwin and F.W. McLafferty, Org. Mass Spectrom., 7 (1973) 1111–1112.
13 P.A. Arpino, B.G. Dawkins and F.W. McLafferty, J. Chromatogr. Sci., 12 (1974) 574–578.
14 F.W. McLafferty and B.G. Dawkins, Biochem. Soc. Trans., 3 (1975) 856–858.
15 B.G. Dawkins and F.W. McLafferty, in Tsuji-Morozowich (Editor), GLC and HPLC Determination of Therapeutic Agents, Vol. 1, Marcel Dekker, New York, 1978, pp. 259–276.
16 P.J. Arpino and G. Guiochon, J. Chromatogr., 251 (1982) 153–164.
17 G. Guiochon and P.J. Arpino, J. Chromatogr., 271 (1983) 13–25.
18 M. Dedieu, C. Juin, P.J. Arpino, J.P. Bounine and G. Guiochon, J. Chromatogr., 251 (1982) 203–213.
19 P.J. Arpino, in T.M. Vickrey (Editor), Liquid Chromatography Detectors, Marcel Dekker, New York, 1983, pp. 243–322.
20 P.J. Arpino, J.P. Bounine and G. Guiochon, J. Chromatogr., 251 (1982) 203–213.
21 P.J. Arpino, J.P. Bounine, M. Dedieu and G. Guiochon, J. Chromatogr., 271 (1983) 43–50.
22 D. Pilosof, H.Y. Kim, D.F. Dyckes and M.L. Vestal, Anal. Chem., in press.
23 C.R. Blakley, J.C. Carmody and M.L. Vestal, Clin. Chem., 26 (1980) 1467–1473.
24 C.R. Blakley and M.L. Vestal, Anal. Chem., 55 (1983) 750–754.
25 M.L. Vestal, Mass Spectrom. Rev., 2 (1983) 447–480.
26 D. Pilosof, H.Y. Kim, M.L. Vestal and D.F. Dyckes, Biomed. Mass Spectrom., in press.
27 H.E. May, F.S. Tanzer, G.H. Fridland, C. Wakelyn and D.M. Desiderio, J. Liq. Chromatogr., 5 (1982) 2135–2154.
28 A.P. Bruins and B.F.H. Drenth, J. Chromatogr., 271 (1983) 71–82.
29 J.D. Henion and G.A. Maylin, Biomed. Mass Spectrom., 7 (1980) 115–121.
30 J.D. Henion, Adv. Mass Spectrom., 8 (1980) 1241–1250.
31 J.D. Henion, J. Chromatogr. Sci., 19 (1981) 57–64.
32 J.D. Henion, B.A. Thomson and P.H. Dawson, Anal. Chem., 54 (1982) 451–456.
33 C.E. Parker, Y. Tondeur, J.R. Hass and J. Forens. Sci., 27 (1982) 495–505.
34 C. Ekers, D.S. Skrabalak and J. Henion, Clin. Chem., 28 (1982) 1882–1886.
35 F.R. Sugnaux, D.S. Skrabalak and J.D. Henion, J. Chromatogr., 264 (1983) 357–376.
36 J. Henion, D. Skrabalak, E. Dewey and G. Maylin, Drug Metabol. Rev., 14 (1983) 961–1003.
37 J. Henion and D.S. Skrabalak, in G.H. Johnston and J.W. Martin (Editors), Fifth International Symposium Equine Medication Control, in press.
38 T. Covey and J. Henion, Anal. Chem., 55 (1983) 2275–2280.
39 R.G. Christensen, H.S. Hertz, S. Meiselman and E. White, Anal. Chem., 53 (1981) 171–174.
40 E. White, H.S. Hertz and R.G. Christensen, U.S. Patent #4281246 (1981).
41 J.A. Apffel, U.A. Th. Brinkman and R.W. Frei, Anal. Chem., 55 (1983) 2280–2284.
42 R.D. Smith and H.R. Udseth, Anal. Chem., 55 (1983) 2266–2272.
43 R.D. Smith, W.D. Feliz, J.C. Fjeldsted and M.L. Lee, Anal. Chem., (1982) 1883–1885.

44   D.J. Liberato, C.C. Fenselau, M.L. Vestal and A.L. Yergey, Anal. Chem., 55 (1983) 1741-1744.
45   C.N. Kenyon, Biomed. Mass Spectrom., 19 (1983) 535-543.
46   B.L. Karger, D.P. Kirby, P. Vouros, R.L. Foltz and B. Hidy, Anal. Chem., 51 (1979) 2324-2328.
47   D.P. Kirby, P. Vouros and B.L. Karger, Science, 209 (1980) 495-497.
48   M.J. Hayes, E.P. Lankmayer, P. Vouros, B.L. Karger and J.M. McGuire, Anal. Chem., 55 (1983) 1745-1752.
49   R.D. Smith and A.L. Johnson, Anal. Chem., 53 (1981) 739-740.
50   T.J. Yu, H. Schwartz, R.W. Giese, B.L. Karger and P. Vouros, J. Chromatogr., 218 (1981) 519-533.
51   T.J. Yu, B.L. Karger and P. Vouros, Biomed. Mass Spectrom., 10 (1983) 633-640.
52   P. Roepstorff, M.A. McDowall, M.P.L. Games and D.E. Games, Int. J. Mass Spectrom. Ion Phys., 48 (1983) 197-200.
53   D.E. Games, M.J. Hewlins, S.A. Westwood and D.J. Morgan, J. Chromatogr., 250 (1982) 62-67.
54   N.J. Alcock, C. Eckers, D.E. Games, M.P.L. Games, M.S. Lant, M.A. McDowall, M. Rossiter, R.W. Smith, S.A. Westwood and H.Y. Wong, J. Chromatogr., 251 (1982) 165-174.
55   D.E. Games, M.A. McDowall, K. Levsen, K.H. Shafer, P. Dobberstein and J.L. Gower, Biomed. Mass Spectrom., 11 (1984) 87-95.
56   R.D. Smith, J.E. Burger and A.L. Johnson, Anal. Chem., 53 (1981) 1603-1611.
57   R.D. Smith and A.L. Johnson, Anal. Chem., 53 (1981) 1120-1122.
58   N.M.M. Nibbering, J. Chromatogr., 251 (1982) 93-104.

SUBJECT INDEX